REINVENTING DISCOVERY

REINVENTING DISCOVERY

The New Era of Networked Science

Michael Nielsen

PRINCETON UNIVERSITY PRESS

PRINCETON AND OXFORD

Published by Princeton University Press, 41 William Street, Princeton, New Jersey 08540
In the United Kingdom: Princeton University Press, 6 Oxford Street, Woodstock,
Oxfordshire OX20 1TW
press.princeton.edu

Library of Congress Cataloging-in-Publication Data

Nielsen, Michael A., 1974–
Reinventing discovery : the new era of networked science / Michael Nielsen.
    p. cm.
Includes bibliographical references and index.
ISBN 978-0-691-14890-8 (hardback)
1. Research–Technological innovations.   2. Discoveries in science.   3. Internet.
4. Information networks.   5. Information technology.   I. Title.
Q180.55.M4N54 2011
509–dc23   2011020248

British Library Cataloging-in-Publication Data is available

This book has been composed in Minion
Printed on acid-free paper ∞

Typeset by S R Nova Pvt Ltd, Bangalore, India
Printed in the United States of America

10 9 8 7 6 5 4 3 2 1

Contents

REINVENTING DISCOVERY

CHAPTER 1

Reinventing Discovery

Tim Gowers is not your typical blogger. A mathematician at Cambridge University, Gowers is a recipient of the highest honor in mathematics, the Fields Medal, often called the Nobel Prize of mathematics. His blog radiates mathematical ideas and insight.

In January 2009, Gowers decided to use his blog to run a very unusual social experiment. He picked out an important and difficult unsolved mathematical problem, a problem he said he'd "love to solve." But instead of attacking the problem on his own, or with a few close colleagues, he decided to attack the problem completely in the open, using his blog to post ideas and partial progress. What's more, he issued an open invitation asking other people to help out. Anyone could follow along and, if they had an idea, explain it in the comments section of the blog. Gowers hoped that many minds would be more powerful than one, that they would stimulate each other with different expertise and perspectives, and collectively make easy work of his hard mathematical problem. He dubbed the experiment the Polymath Project.

The Polymath Project got off to a slow start. Seven hours after Gowers opened up his blog for mathematical discussion, not a single person had commented. Then a mathematician named Jozsef Solymosi from the University of British Columbia posted a comment suggesting a variation on Gowers's problem, a variation which was easier, but which Solymosi thought might throw light on the original problem. Fifteen minutes later, an Arizona high-school teacher named Jason Dyer chimed in with a thought of his own. And just three minutes after that, UCLA mathematician Terence Tao—like Gowers, a Fields medalist—added a comment. The comments erupted: over the next 37 days, 27 people wrote 800 mathematical comments, containing more than 170,000 words. Reading through the comments you see ideas proposed, refined, and discarded, all with incredible speed. You see top mathematicians making mistakes, going down wrong paths, getting their hands dirty following up the most mundane of details, relentlessly pursuing a solution. And through all the false starts and wrong turns, you see a gradual dawning of insight. Gowers described the Polymath process as being "to normal research as driving is to pushing a car." Just 37 days after the project began Gowers announced that he was confident the polymaths had solved not just his original problem, but a harder problem that included the original as a special case. He described it as "one of the most exciting six weeks of my mathematical life." Months' more cleanup work remained to be done, but the core mathematical problem had been solved. (If you'd like to know the details of Gowers's problem, they're described in the appendix. If you just want to get on with reading this book, you can safely skip those details.)

The polymaths aren't standing still. Since Gowers's original project, nearly a dozen Polymath and Polymath-like projects have been launched, some attacking problems even more ambitious than Gowers's original problem. More than 100 mathematicians and other scientists have participated; mass collaboration is starting to spread through mathematics. Like the first Polymath Project, several of these projects have been great successes, really driving our under-standing of mathematics forward. Others have been more qualified successes, falling short of achieving their (sometimes extremely ambitious) goals. Regardless, massively collaborative mathematics is a powerful new way of attacking hard mathematical problems.

Why is mass online collaboration useful in solving mathematical problems? Part of the answer is that even the best mathematicians can learn a great deal from people with complementary knowledge, and be stimulated to consider ideas in directions they wouldn't have considered on their own. Online tools create a shared space where this can happen, a short-term collective working memory where ideas can be rapidly improved by many minds. These tools enable us to scale up creative conversation, so connections that would ordinarily require fortuitous serendipity instead happen as a matter of course. This speeds up the problem-solving process, and expands the range of problems that can be solved by the human mind.

The Polymath Project is a small part of a much bigger story, a story about how online tools are transforming the way scientists make discoveries. These tools are *cognitive tools*, actively amplifying our collective intelligence, making us smarter and so better able to solve the toughest scientific problems. To understand why all this matters, think back to the seventeenth century and the early days of modern science, the time of great discoveries such as Galileo's observation of the moons of Jupiter, and Newton's formulation of his laws of gravitation. The greatest legacy of Galileo, Newton, and their contemporaries wasn't those one-off breakthroughs. It was the method of scientific discovery itself, a way of understanding how nature works. At the beginning of the seventeenth century extraordinary genius was required to make even the tiniest of scientific advances. By developing the method of scientific discovery, early scientists ensured that by the end of the seventeenth century such scientific advances were run-of-the-mill, the likely outcome of any competent scientific investigation. What previously required genius became routine, and science exploded.

Such improvements to the way discoveries are made are more important than any single discovery. They extend the reach of the human mind into new realms of nature. Today, online tools offer us a fresh opportunity to improve the way discoveries are made, an opportunity on a scale not seen since the early days of modern science. I believe that the process of science—how discoveries are made—will change more in the next twenty years than it has in the past 300 years.

The Polymath Project illustrates just a single aspect of this change, a shift in how scientists work together to create knowledge. A second aspect of this change is a dramatic expansion in scientists' ability to find meaning in knowledge. Consider, for example, the studies you often see reported in the news saying "so-and-so genes cause such-and-such a disease." What makes these studies possible is a genetic map of human beings that's been assembled over the past twenty years. The best-known part of that map is the human genome, which scientists completed in 2003. Less well known, but perhaps even more important, is the HapMap (short for haplotype map), completed in 2007, which charts how and where different human beings can *differ* in their genetic code. Those genetic variations determine much about our different susceptibilities to disease, and the HapMap says where those variations can occur—it's a genetic map not just of a single human being, but of the entire human race.

This human genetic map was the combined work of many, many biologists around the world. Each time they obtained a new chunk of genetic data in their laboratories, they uploaded that data to centralized online services such as GenBank, the amazing online repository of genetic information run by the US National Center for Biotechnology Information. GenBank integrates all this genetic information into a single, publicly accessible online database, a compilation of the work of thousands of biologists. It's information on a scale that's almost impossible to analyze by hand. Fortunately, anyone in the world may freely download the genetic map, and then use computer algorithms to analyze the map, perhaps discovering previously unsuspected facts about the human genome. You can, if you like, go to the GenBank site right now, and start browsing genetic information. (For links to GenBank and other resources, see the "Notes on Sources," starting on page 347.) This is, in fact, what makes those studies linking genes to disease possible: the scientists doing the studies start by finding a large group of people with the disease, and also a control group of people without the disease. They then use the human genetic map to find correlations between disease incidence and the genetic differences of the two groups.

A similar pattern of discovery is being used across science. Scientists in many fields are collaborating online to create enormous databases that map out the structure of the universe, the world's climate, the world's oceans, human languages, and even all the species of life. By integrating the work of hundreds or thousands of scientists, we are collectively mapping out the entire world. With these integrated maps anyone can use computer algorithms to discover connections that were never before suspected. Later in the book we'll see examples ranging from new ways of tracking influenza outbreaks to the discovery of orbiting pairs of supermassive black holes. We are, piece by piece, assembling all the world's knowledge into a single giant edifice. That edifice is too vast to be comprehended by any individual working alone. But new computerized tools can help us find meaning hidden in all that knowledge.

If the Polymath Project illustrates a shift in how scientists collaborate to create knowledge, and GenBank and the genetic studies illustrate a shift in how scientists find meaning in knowledge, a third big shift is a change in the relationship between science and society. An example of this shift is the website Galaxy Zoo, which has recruited more than 200,000 online volunteers to help astronomers classify galaxy images. Those volunteers are shown photographs of galaxies, and asked to answer questions such as "Is this a spiral or an elliptical galaxy?" and "If this is a spiral, do the arms rotate clockwise or anticlockwise?" These are photographs that have been taken automatically by a robotic telescope, and have never before been seen by any human eye. You can think of Galaxy Zoo as a cosmological census, the largest ever undertaken, a census that has so far produced more than 150 million galaxy classifications.

The volunteer astronomers who participate in Galaxy Zoo are making astonishing discoveries. They have, for example, recently discovered an entirely new class of galaxy, the "green pea galaxies"— so named because the galaxies do, indeed, look like small green peas—where stars are forming faster than almost anywhere else in the universe. They've also discovered what is believed to be the first ever example of a quasar mirror, an enormous cloud of gas tens of thousands of light-years in diameter, which is glowing brightly as the gas is heated by light from a nearby quasar. In just three years, the work of the Galaxy Zoo volunteers has resulted in 22 scientific papers, and many more are in the works.

Galaxy Zoo is just one of many online citizen science projects that are recruiting volunteers, most of them without scientific training, to help solve scientific research problems. We'll see examples ranging across science, from volunteers who are using computer games to predict the shape of protein molecules, to volunteers who are helping understand how dinosaurs evolved. These are serious scientific projects, projects where large groups of volunteers with little scientific training can attack scientific problems beyond the reach of small groups of professionals. There's no way a team of professionals could do what Galaxy Zoo does—even working full time, the pros don't have the time to classify hundreds of thousands (or more) of galaxies. You might suppose they'd use computers to classify the galaxy images, but in fact the human volunteers classify the galaxies more accurately than even the best computer programs. So the volunteers at projects such as Galaxy Zoo are expanding the boundary of what scientific problems can be solved, and in so doing, changing both who can be a scientist and what it means to be a scientist. How far can the boundary between professional and amateur scientist be blurred? Will we one day see Nobel Prizes won by huge collaborations dominated by amateurs?

Citizen science is part of a larger shift in the relationship between science and society. Galaxy Zoo and similar projects are examples of institutions that are bridging the scientific community and the rest of society in new ways. We'll see that online tools enable many other new bridging institutions, including open access publishing, which gives the public direct access to the results of science, and science blogging, which is helping create a more open and more transparent scientific community. What other new ways can we find to build bridges between science and the rest of society? And what will be the long-run impact of these new bridging institutions?

The story so far is an optimistic story of possibility, of new tools that are changing the world. But there's a problem with this story, some major obstacles that prevent scientists from taking full advantage of online tools. To understand the obstacles, consider the studies linking genes to disease that we discussed earlier. There's a crucial part of that story which I glossed over, but which is actually quite puzzling: *why* is it that biologists share genetic data in GenBank in the first place? When you think about it, it's a peculiar choice: if you're a professional biologist it's to your advantage to keep data secret as long as possible. Why share your data online before you get a chance to publish a paper or take out a patent on your work? In the scientific world it's papers and, in some fields, patents that are rewarded by jobs and promotions. Publicly releasing data typically does nothing for your career, and might even damage it, by helping your scientific competitors.

In part for these reasons, GenBank took off slowly after it was launched in 1982. While many biologists were happy to access others' data in GenBank, they had little interest in contributing their own data. But that has changed over time. Part of the reason for the change was a historic conference held in Bermuda in 1996, and attended by many of the world's leading biologists, including several of the leaders of the government-sponsored Human Genome Project. Also present was Craig Venter, who would later lead a private effort to sequence the human genome. Although many attendees weren't willing to unilaterally make the first move to share all their genetic data in advance of publication, everyone could see that science as a whole would benefit enormously if open sharing of data became common practice. So they sat and talked the issue over for days, eventually coming to a joint agreement— now known as the Bermuda Agreement—that all human genetic data should be immediately shared online. The agreement wasn't just empty rhetoric. The biologists in the room had enough clout that they convinced several major scientific grant agencies to make immediate data sharing a mandatory requirement of working on the human genome. Scientists who refused to share data would get no grant money to do research. This changed the game, and immediate sharing of human genetic data became the norm. The Bermuda agreement eventually made its way to the highest levels of government: on March 14, 2000, US President Bill Clinton and UK Prime Minister Tony Blair issued a joint statement praising the principles described in the Bermuda Agreement, and urging scientists in every country to adopt similar principles. It's because of the Bermuda Agreement and similar subsequent agreements that the human genome and the HapMap are publicly available.

This is a happy story, but it has an unhappy coda. The Bermuda Agreement originally only applied to human genetic data. There have since been many attempts to extend the spirit of the agreement, so that more genetic data is shared. But despite these attempts, there are still many forms of life for which genetic data remains secret. For example, as of 2010 there is no worldwide agreement to share data about the influenza virus. Steps toward such an agreement remain bogged down in wrangling among the leading parties. To give you the flavor of how many scientists think about sharing non-human genetic data, one scientist recently told me that he'd been "sitting on a genome" for an entire species (!) for more than a year. Without any incentive to share, and with many reasons not to, scientists hoard their data. As a result, there's an emerging data divide between our understanding of life-forms such as human beings, where nearly all genetic data are available online, and life-forms such as influenza, where important data remain locked up.

This story makes it sound as though the scientists involved are greedy and destructive. After all, this research is typically paid for using public funds. Shouldn't scientists make their results available as soon as possible? There's truth to these ideas, but the situation is complex. To understand what's going on, you need to understand the incredible competitive pressures on ambitious young scientists. On the rare occasion a good long-term job at a major university opens up, there are often hundreds of superbly-qualified applicants. Competition for jobs is so fierce that eighty-hour-plus workweeks are common among young scientists. As much of that time as possible is spent working on the one thing that will get such a job: amassing an impressive record of scientific papers. Those papers will bring in the research grants and letters of recommendation necessary to find long-term employment. The pace relaxes after tenure, but continued grant support still requires a strong work ethic. The result is that while many scientists agree in principle that they'd love to share their data in advance of publication, they worry that doing so will give their competitors an unfair advantage. Those competitors could exploit that knowledge to rush their results into print first, or, worse, even steal the data outright and present the results as their own. It's only practical to share data if everyone is protected by a collective agreement such as the Bermuda agreement.

A similar pattern has seen scientists resist contributing to many other online projects. Consider Wikipedia, the online encyclopedia. Wikipedia has a vision statement to warm a scientist's heart: *"Imagine a world in which every single human being can freely share in the sum of all knowledge. That's our commitment."* You might think Wikipedia was started by scientists eager to share all the world's knowledge, but you'd be wrong. In fact, it was started by Jimmy "Jimbo" Wales, who at the time was cofounder of an online company mostly specializing in adult content, and Larry Sanger, a philosopher who left academia to work with Wales on online encyclopedias. In the early days of Wikipedia there was little involvement from scientists. This was despite the fact that anyone in the world can edit Wikipedia, and, in fact, it's written entirely by its users. So here's this incredibly exciting project, which anyone can get involved in, which is taking off rapidly, and which expresses core scientific values. Why weren't scientists lining up to be involved? The problem is the same as with the genetic data: why would scientists take the time to contribute to Wikipedia when they could be doing something more respectable among their peers, like writing a paper? That's the kind of activity that leads to jobs, grants, and promotions. It doesn't matter that contributing to Wikipedia might be more intrinsically valuable. In the early days work on Wikipedia was seen by scientists as frivolous, a waste of time, as not being serious science. I'm happy to say that this has changed over the years, and today Wikipedia's success has to some extent legitimized work on it by scientists. But isn't it strange that the modern-day Library of Alexandria came from outside academia?

There's a puzzle here. Scientists helped create the internet and the world wide web. They've taken enthusiastically to online tools such as email, and pioneered striking projects such as the Polymath Project and Galaxy Zoo. Why is it that they've only reluctantly adopted tools such as GenBank and Wikipedia? The reason is that, despite their radical appearance, the Polymath Project, Galaxy Zoo, and similar undertakings have an inherent underlying conservatism: they're ultimately projects in service of the conventional goal of writing scientific papers. That conservatism helps them attract contributors who are willing to use unconventional means such as blogs to more effectively achieve a conventional end (writing a scientific paper). But when the goal isn't simply to produce a scientific paper —as with GenBank, Wikipedia, and many other tools—there's no direct motivation for scientists to contribute. And that's a problem, because some of the best ideas for improving the way scientists work involve a break away from the scientific paper as the ultimate goal of scientific research. There are opportunities being missed that dwarf GenBank and Wikipedia in their potential impact. In this book, we'll delve into the history and culture of science, and see how this situation arose, in which scientists are often reluctant to share their ideas and data in ways that speed up the advancement of science. The good news is that we'll find leverage points where small changes today will lead to a future where scientists do take full advantage of online tools, greatly increasing our capacity for scientific discovery.

Revolutions are sometimes marked by a single, spectacular event: the storming of the Bastille during the French Revolution, or the signing of the US Declaration of Independence. But often the most important revolutions aren't announced with the blare of trumpets. They occur quietly, too slowly to make the news, but fast enough that if you aren't alert, the revolution is over before you're aware it's happening. The change described in this book is like this. It's not a single event, nor is it a change that's happening quickly. It's a slow revolution that has quietly been gathering steam for years. Indeed, it's a change that many scientists have missed or underestimated, being so focused on their own specialty that they don't appreciate just how broad-ranging the impact of the new online tools is. They're like surfers at the beach who are so intent on watching the waves crash and recede that they're missing the rise of the tide. But you shouldn't let the slow, quiet nature of the current changes in how science is done fool you. We are in the midst of a great change in how knowledge is constructed. Imagine you were alive in the seventeenth century, at the dawn of modern science. Most people alive at that time had no idea of the great transformation that was going on, a transformation in how we know. Even if you were not a scientist, wouldn't you have wanted to at least be aware of the remarkable transformation that was going on in how we understood the world? A change of similar magnitude is going on today: we are reinventing discovery.

I wrote this book because I believe the reinvention of discovery is one of the great changes of our time. To historians looking back a hundred years from now, there will be two eras of science: pre-network science, and networked science. We are living in the time of transition to the second era of science. But it's going to be a bumpy transition, and there is a possibility it will fail or fall short of its potential. And so I also wrote the book to help create a widely shared public understanding of the opportunity now before us, an understanding that a more open approach to science isn't just a nice idea, but that it must be demanded of our scientists and our scientific institutions.

This change is important. Improving the way science is done means speeding up the rate of all scientific discovery. It means speeding up things such as curing cancer, solving the climate-change problem, launching humanity permanently into space. It means fundamental insights into the human condition, into how the universe works and what it is made of. It means discoveries we've not yet dreamt of. Over the next few years we have an astonishing opportunity to change and improve the way science is done. This book is the story of this change, what it means for us, and what we need to do to make it happen.

PART 1

Amplifying Collective Intelligence

CHAPTER 2

Online Tools Make Us Smarter

In 1999, world chess champion Garry Kasparov played a game of chess against "the World." In this event, organized by Microsoft, the idea was that anyone in the world could go to the game website, and vote on what move should be taken next. On a typical move more than 5,000 people voted, and over the entire game 50,000 people from 75 countries voted. The World Team decided on a new move every 24 hours, and on any given turn the move taken was whichever got the most votes. The game was billed as "Kasparov versus the World."

The game exceeded all expectations. After 62 moves of innovative chess, in which the balance of the game changed several times, the World Team finally resigned. Kasparov called it "the greatest game in the history of chess," and revealed that during the game he often couldn't tell who was winning and who was losing; it wasn't until the 51st move that the balance swung decisively in his favor. After the game, Kasparov wrote a book about it, and in that book he commented that he expended more energy on this one game than on any other in his career, including world championship games.

Although the World Team had input from some strong players, none was as strong as Kasparov himself, and the average quality of player was far below Kasparov. Yet collectively the World Team played a game far stronger than any of the individuals contributing would ordinarily have played—indeed, one of the strongest games in the history of chess. Not only did they play Kasparov at his best, but much of their deliberation about strategy and tactics was carried out in public, an advantage Kasparov used extensively. Imagine not only playing Garry Kasparov, but also having to explain to him the thinking behind your moves!

How was this possible? How could thousands of chess players, most of them amateurs, compete in a chess game with Kasparov at his peak? The World Team contained people at all levels of chess ability, from beginners to grandmasters. Moves regarded by experts as obviously mistaken sometimes obtained up to 10 percent of the vote, suggesting that many beginners were participating. On one move, 2.4 percent of the votes were cast for moves that weren't merely bad, but actually violated the rules of chess!

The World Team coordinated their play in several ways. Microsoft set up a game forum where people could discuss the game, and also appointed four official advisors to the World Team. These were outstanding teenage chess players, among the best of their age in the world, although none were in Kasparov's class. On each move, the advisors published their recommendations on the Microsoft website, and, if they wanted, a commentary explaining the recommendation. This was done well before World Team voting closed, so the recommendations could influence the voting. As the game progressed several other strong chess players also offered advice. Particularly influential, although not always heeded, was the GM School, a Russian chess club containing several grandmasters.

Most of the advisors and other strong players ignored the discussion on the game forum, making no attempt to engage with the bulk of people making up the World Team, and so distancing themselves from the people whose votes decided the World Team's moves. But one of the advisors did actively engage with the World Team. This was an extraordinary young chess player named Irina Krush. Fifteen years old, Krush had recently become the US Women's chess champion. Although not as highly rated as two of the other World Team advisors, Krush was certainly in the international elite of junior chess players.

Unlike her expert peers, Krush devoted considerable time and attention to the World Team's game forum. Shrugging off abuse and insults, she extracted many of the best ideas and analyses from the forum, wrote extensive commentary describing the thinking behind her recommended moves, and gradually built up a network of strong chess-playing correspondents, including some of the grandmasters offering advice.

Simultaneously, Krush and her management team, a company named Smart Chess, built a publicly accessible analysis tree for the game, showing possible moves and countermoves, and containing the best arguments for and against different lines of play. These arguments were drawn not only from her own analysis, but also from the game forum and from her correspondence with others, including the GM School. This analysis tree helped the World Team coordinate their efforts, prevented duplication of effort, and served as a reference point for the World Team during discussion and voting.

As the game went on, Krush's role on the World Team became pivotal. Part of the reason was the quality of her play. On move 10, Krush suggested a move that Kasparov called "a great move, an important contribution to chess," blowing the game wide open, and taking it into uncharted chess territory. This move raised her standing with the World Team, and helped her assume a coordinating role. Between moves 10 and 50 Krush's recommended move was always played by the World Team, even when it disagreed with the recommendations of the other three advisors to the World Team, or with influential commentators such as the GM School.

As a result, some people say the game was really Kasparov versus Krush, despite the fact that Kasparov would ordinarily have beaten Krush easily. Kasparov himself has said he believed he was really playing against Smart Chess, Krush's management team. Krush has dismissed both points of view. In a series of essays written after the game she explained the thinking behind her recommended moves, and how she drew on ideas from a multitude of sources, ranging from anonymous posters on the game forum to grandmasters. She repeatedly explains how she changed and in some cases abandoned her own ideas, convinced by someone else's superior analysis. Thus, Krush was neither playing alone nor as part of a small team, but rather was at the center of the coordination effort for the entire World Team. As a result she had the best understanding of all the suggestions being made by members of the World Team. Other, stronger players didn't understand the different points of view as well, and so didn't make as good decisions about what move to make next, nor did they have the standing with the World Team to influence the voting as strongly as Krush. Krush's coordinating role thus brought the best ideas of all contributors into a coherent whole. The result was that the World Team emerged far stronger than any individual player on the team, and arguably stronger than any player in history except Kasparov at his peak.

Kasparov versus the World was not the first game to pit a chess grandmaster against the World. Three years earlier, in 1996, former world chess champion Anatoly Karpov also played such a game. "Karpov Against the World" used a different online system to decide moves, with no game forum or official game advisors, and giving World Team members just ten minutes to vote on their preferred move. Without the means to coordinate their actions, the World Team played poorly, and Karpov crushed them in just 32 moves. Perhaps influenced by Karpov's success, Kasparov admitted that before his game he "was not anticipating any particular difficulties," and was confident that he "would be able to finish matters in under 40 moves." How surprised he must have been.

Amplifying Collective Intelligence

Examples such as Kasparov versus the World and the Polymath Project show that groups can use online tools to make themselves collectively smarter. That is, those tools can be used to amplify our collective intelligence, in much the way that manual tools have been used for millennia to amplify our physical strength. How do these new tools achieve this amazing feat? Is it just a fluke? Or can online tools be used more generally to solve creative problems that defeat the ingenuity of even the cleverest individuals? Are there general design principles that can be used to amplify collective intelligence, a sort of design science of collaboration?

A common approach to these questions is to suggest that online tools enable some sort of collective brain, with the people in the group playing the role of neurons. A greater intelligence then somehow emerges from the connections between these human neurons. While this metaphor is stimulating, it has many problems. The brain's origin and hardware are completely different from those of the internet, and there's no compelling reason to suppose the brain is an accurate model of how collective intelligence works, or of how it can best be amplified. Whatever our collective brain is doing, it seems likely to work according to very different principles than the brain inside our heads. Furthermore, we don't yet have a good understanding of how the human brain works, so the metaphor is in any case of limited use at best. If we're going to understand how to amplify collective intelligence, we need to look beyond the metaphor of the collective brain.

Many books and magazine articles have been written about collective intelligence. Perhaps the best-known example of this work is James Surowiecki's 2004 book *The Wisdom of Crowds*, which explains how large groups of people can sometimes perform surprisingly well at problem solving. Surowiecki opens his book with a striking story about the scientist Francis Galton. In 1906, Galton was attending an English country fair, and among the fair's attractions was a weight-judging contest, where people competed to guess the weight of an ox. Galton expected that most of the competitors would be far off in their estimates, and was surprised to learn that the average of all the competitors' guesses (1,197 pounds) was just one pound short of the correct weight of 1,198 pounds. In other words, collectively, if one averaged the guesses, the crowd at the fair guessed the weight almost perfectly. Surowiecki's book goes on to discuss many other ways we can combine our collective wisdom to surprisingly good effect.

This book goes beyond *The Wisdom of Crowds* and similar works in two ways. First, our goal is to understand how online tools can actively *amplify* collective intelligence. That is, we're not just interested in collective intelligence, per se, but in how to design tools that dramatically increase collective intelligence. Second, we're not just discussing everyday problems like estimating the weight of an ox. Instead, our focus is on problems at the limit of human problem-solving ability, problems like competing with Garry Kasparov at the peak of his chess-playing might, or smashing mathematical problems that challenge the world's best mathematicians. Our main interest will be in scientific problem-solving, and of course it's problems at the limit of human problem-solving ability that scientists most dearly want to solve, and whose solution will bring the greatest benefit.

Superficially, the idea that online tools can make us collectively smarter contradicts the idea, currently fashionable in some circles, that the internet is reducing our intelligence. For example, in 2010 the author Nicholas Carr published a book entitled *The Shallows: What the Internet Is Doing to Our Brains*, arguing that the internet is reducing our ability to concentrate and contemplate. Carr's book and other similar works make many good points, and have been widely discussed. But new technologies seldom have just a single impact, and there's no contradiction in believing that online tools can both enhance and reduce intelligence. You can use a hammer to build a house; you can also use it to break your thumb. Complex technologies, especially, often require considerable skill to use well. Automobiles are amazing tools, but we all know how learner drivers can terrorize the road. Looking at the internet and concluding that the main impact is to make us stupid is like looking at the automobile and concluding that it's a tool for learner drivers to wipe out terrified pedestrians. Online, we're all still learner drivers, and it's not surprising that online tools are sometimes used poorly, amplifying our individual and collective stupidity. But as we've already seen, there are also examples showing that online tools can be used to increase our collective intelligence. Our concern will therefore be with understanding how those tools can be used to make us collectively smarter, and what that change will mean.

We're still in the early days of understanding how to amplify collective intelligence. It's telling that many of the best tools we have—tools such as blogs, wikis, and online forums—weren't invented by the people we might suppose are the experts on group behavior and intelligence, experts from fields such as group psychology, sociology, and economics. Instead, they were invented by amateurs, people such as Matt Mullenweg, who was a 19-year-old student when he created Wordpress, one of the most popular types of blogging software, and Linus Torvalds, who was a 21-year-old student when he created the open source Linux operating system. That tells us we should be wary of current theory: while we can learn a great deal from existing academic studies, the picture of collective intelligence that emerges is also incomplete. For this reason, we'll ground our discussion in concrete examples in the mold of the Polymath Project and Kasparov versus the World. In part 1 of this book we'll use these concrete examples to distill a set of principles that explain how online tools can amplify collective intelligence.

I have deliberately focused the discussion in part 1 on a relatively small number of examples, with the idea being that as we develop a conceptual framework for understanding collective intelligence, we'll revisit each of these examples several times, and come to understand them more deeply. Furthermore, the examples come not just from science, but also from areas such as chess and computer programming. The reason is that some of the most striking examples of amplifying collective intelligence—examples such as Kasparov versus the World—come from outside science, and we can learn a great deal by studying them.

As our understanding deepens, we'll see that scientific problems are especially well suited for attack by collective intelligence, and in part 2 we'll narrow our focus to how collective intelligence is changing science.

CHAPTER 3

Restructuring Expert Attention

In 2003, a young woman named Nita Umashankar, from Tucson, Arizona, went to live for a year in India, where she worked with a not-for-profit organization to help sex workers escape the sex trade. What she found in India frustrated her. Many of the sex workers had so few skills that it was almost impossible to help them find jobs outside prostitution. Returning to the United States, Umashankar decided she would start a not-for-profit organization that addressed the core problem, by training at-risk Indian girls in technology, and then helping them find jobs with technology companies.

Eight years later, the nonprofit she founded, ASSET India, has opened technology training centers in five Indian cities. They've helped hundreds of young people escape the sex trade, and have plans to expand. Unfortunately, many of the smaller towns they'd like to expand into don't have the reliable electricity needed to power crucial technologies such as the wireless routers used to access the internet. ASSET has experimented with using solar-powered wireless routers, but found that the devices already on the market won't run reliably over the long hours their training centers are open.

To solve their problem with wireless routers, ASSET tried something unconventional, searching for help using an online marketplace for scientific problems called InnoCentive. InnoCentive is like eBay or Craigslist, but aimed at scientific problems. The idea is that participating organizations can post online "Challenges"— scientific problems they want solved—with prizes for solution, often tens of thousands of dollars. Anyone in the world can download a detailed description of a Challenge, try to solve the problem, and win the prize.

Using $20,000 in prize money put up by the Rockefeller Foundation, ASSET posted an InnoCentive Challenge to design a reliable solar-powered wireless router, using low-cost, easily available hardware and software. In the two months the Challenge was posted at InnoCentive it was downloaded 400 times, and 27 solutions were submitted. The $20,000 prize was awarded to a 31-year-old Texan software engineer named Zacary Brown, and a prototype is being built by engineering students at the University of Arizona.

Zacary Brown wasn't just any software engineer. An enthusiastic amateur wireless radio operator, he was working toward a goal of making radio contact with every country in the world. While growing up, he was enchanted by his parents' explanation of how the solar panels Jimmy Carter installed at the White House made electricity from sunlight, and as an adult he was experimenting with using solar panels to power his wireless radio equipment. Over the long run, he hoped to power his entire home office using solar power. In short, Zacary Brown was exactly the right person for ASSET to be talking to. InnoCentive simply provided a way of making the connection.

Underlying InnoCentive is the premise that there is enormous untapped potential for scientific discovery in the world, potential that can be released by connecting the right people. This premise has been confirmed, with more than 160,000 people from 175 countries signing up to InnoCentive, and prizes for more than 200 Challenges awarded. The Challenges range across many areas of science and technology. Examples include finding more cost-effective methods of manufacturing drugs for tuberculosis, designing a solar-powered mosquito repellent (I'm not making this up!) to combat malaria, and finding better ways of identifying people at risk of developing motor neuron disease. Many of the successful solvers report, as Zacary Brown did, that the Challenges they solve closely match their skills and interests. Furthermore, as in the ASSET story, connections are usually made between parties who otherwise would only have met accidentally. InnoCentive makes such connections systematically, not as lucky one-offs, but at scale.

The reason the connections made by InnoCentive are so valuable is, of course, the big gap between the skills of the people posing the Challenges and those solving the Challenges. While designing a solar-powered wireless router may take an expert such as Zacary Brown only a few days, it would take months or years for the people at ASSET India. They just don't have the right expertise. It's because Zacary Brown has such an enormous comparative advantage that he and ASSET can work together for mutual benefit. More generally, the attention of the right expert at the right time is often the single most valuable resource one can have in creative problem solving. Expert attention is to creative problem solving what water is to life in the desert: it's the fundamental scarce resource. InnoCentive creates value by *restructuring expert attention*, so that people such as Zacary Brown can use their expertise in high-leverage ways: InnoCentive helped Zacary Brown focus his expertise on ASSET's problem, instead of working at home on his hobbies.

In this chapter we'll see that it's this ability to restructure expert attention that is at the heart of how online tools amplify collective intelligence. What examples such as InnoCentive, the Polymath Project, and Kasparov versus the World share is the ability to bring the attention of the right expert to the right problem at the right time. In the first half of the chapter we'll look in more detail at these examples, and develop a broad conceptual framework that explains how they restructure expert attention. In the second half of the chapter we'll apply that framework to understand how online collaborations can work together in ways that are essentially different from offline collaborations.

Harnessing Latent Microexpertise

While the ASSET-InnoCentive story is striking, Kasparov versus the World is an even more impressive example of collective intelligence. As in the ASSET-InnoCentive story, Kasparov versus the World relied on a restructuring of expert attention. To understand how this worked, let's return to the game-making move suggested by Irina Krush, move number 10, the move Kasparov praised as "a great move, an important contribution to chess." Krush's suggestion didn't come from thin air. She had the idea for move 10 a full month before Kasparov versus the World began, during a study session at the World Open chess tournament in Philadelphia. At the time, she did a brief analysis, and talked the move over with her trainers, grandmasters Giorgi Kacheishvili and Ron Henley, before putting the idea aside. It was a lucky chance that Kasparov versus the World opened in a way that let Krush use the move she'd been considering in Philadelphia. It certainly wasn't something she could completely control, because Kasparov was playing the white pieces, and so playing first, which allowed him to dictate the initial direction of the game. Still, a full week before move 10 was played, Krush and her trainers were alive to the possibility that the game might head in this direction, and began to analyze the pros and cons of Krush's idea more intensively.

It's important to appreciate that in nearly all ways Kasparov was far and away Krush's superior as a chess player. We can express the gap between them quite precisely, since there is a numerical rating system that is used to rate chess players. In that rating system a good club player will have a rating in the range of 1,800 to 2,000. An international master such as Irina Krush will have a rating around 2,400. In 1999, at the time of his game against the World, Kasparov's rating peaked at 2,851—not only the highest rating in chess history, but considerably higher than any other player's rating before or since. The 450-point rating gap between Kasparov and Krush was roughly the same as the gap between Krush and a good club player. It meant that Krush would only stand a chance of winning a game against Kasparov if he made a major blunder. This is not to say that Krush was a weak player—remember, she was the U.S. women's champion—but at the time of the game Kasparov was in another class.

Given the large gap in ability between Kasparov and Krush, it appears very fortunate that the game unfolded in a way that gave Krush a chance to exploit her extremely specialized expertise about the opening that led to move 10. In this narrow slice of chess, she was Kasparov's superior, and could give the World Team an advantage. Put another way, although Krush was inferior to Kasparov in nearly all areas of chess, in this special area of *microexpertise* she surpassed even Kasparov.

But although it was luck that Krush's particular microexpertise could help the World Team get the upper hand, that doesn't mean it was simply luck that enabled the World Team to play so well.

The game was widely publicized within the chess community, and hundreds of experienced chess players were following the game. Chess is so rich with possible variations that many of those players had their own individual areas of microexpertise where they too equaled or even surpassed Kasparov. The key to the World Team's play was to ensure that this all this ordinarily latent microexpertise was uncovered and acted upon in response to the contingencies of the game. So although it was a lucky chance that Krush *in particular* was the person whose microexpertise was decisive at move 10, given the number of experienced chess players involved, it was highly likely that latent microexpertise from those players would come to light at critical points during the game, and so help the World Team match Kasparov.

This is, in fact, exactly what happened. As an example, after the game ended Krush singled out move number 26 as one of her three favorite World Team moves. Move number 26 wasn't Krush's idea, or the idea of one of the established chess experts following the game. Instead, move 26 was proposed by one of the posters on the game forum, using the name Yasha, later revealed to be Yaaqov Vaingorten, a reasonably serious but not elite junior player. This was part of a pattern, as during the game Krush drew extensively on the thinking of many unknown or even anonymous contributors to the game forum, people using pseudonyms such as Agent Scully, Solnushka, and Alekhine via Ouji. At the same time, she also consulted with established chess players such as international masters Ken Regan and Antti Pihlajasalo, and grandmaster Alexander Khalifman, of the GM School. The World Team wasn't lucky at all. Rather, the World Team had such a diverse collection of talent available that each time a problem arose, a member of the team rose to the occasion; someone with just the right microexpertise would leap in to fill the gap.

Designed Serendipity

We've seen how collaborative projects such as Kasparov versus the World and InnoCentive harness latent microexpertise to overcome challenges that would stymie most members of the collaboration.

In the most successful online collaborations this use of micro-expertise approaches an ideal in which the collaboration *routinely* locates people such as Yasha and Zacary Brown, people with just the right microexpertise for the occasion. In particular, as creative collaboration is scaled up, problems can be exposed to people with a greater and greater range of expertise, greatly increasing the chance that someone will see what seems to most participants like a hard problem and think, "Hey, that's easy to solve." Instead of being an occasional fortuitous coincidence, serendipity becomes commonplace. The collaboration achieves a kind of *designed serendipity*, a term I've adapted from the author Jon Udell.

To understand the value of such serendipity in creative work, it helps to have a concrete historical example. Let's take Einstein's work on his greatest contribution to science, his theory of gravity, often called the general theory of relativity. He worked on and off developing general relativity between 1907 and 1915, often running into great difficulties. By 1912, his work had led him to the astonishing conclusion that our ordinary conception of the geometry of space, in which the angles of a triangle add up to 180 degrees, is only approximately correct, and a new kind of geometry is needed to describe space and time. Now, in case you're wondering what the geometry of space and time has to do with gravity, you're in good company: it came as a surprise to Einstein, too. When setting out to understand gravity, Einstein had no idea that he'd end up thinking of it as a geometric problem. Nonetheless, there he stood in 1912 with the idea that gravity was somehow connected to a nonstandard type of geometry. And he was stuck, because such geometric ideas were outside his expertise. He talked his problems over with a long-time mathematician friend, Marcel Grossmann, telling him, "Grossmann, you must help me or else I'll go crazy!" Fortunately, for Einstein, Grossmann was just the right person to be talking to. He told Einstein that the geometric ideas Einstein needed had already been worked out in full, decades earlier, by the mathematician Bernhard Riemann. Einstein quickly dove into Riemannian geometry, and realized that Grossmann was right. Riemannian geometry became the mathematical language of general relativity.

Serendipitous connections like this are crucial in creative work. In science, especially, every active scientist carries around in their head a host of unsolved problems. Some of those problems are big ("Figure out how the universe began"), some of them are small ("Where'd that damned minus sign disappear in my calculation?"), but all of them are grist for future progress. If you're a scientist, it's mostly up to you to solve those problems by yourself. If you're lucky, you might have a few supportive colleagues who can help you out. Very occasionally, though, you'll solve a problem in a completely different way. You'll be talking with an acquaintance, when one of your problems comes up. You're chatting away when BANG, all of a sudden you realize that this is exactly the right person to be talking to. Sometimes, they can just outright solve your problem. Or sometimes they give you some crucial insight or idea that provides the momentum needed to vanquish the problem. This kind of fortuitous connection is one of the most exciting and important moments in science. The problem is, such chance connections occur too rarely. The reason designed serendipity is important is because in creative work, most of us—even Einstein!—spend much of our time blocked by problems that would be routine, if only we could find the right expert to help us. As recently as 20 years ago, finding that right expert was likely to be difficult. But, as examples such as InnoCentive and Kasparov versus the World show, we can now design systems that make it routine. Designed serendipity enables us to rapidly and routinely solve many of those previously insoluble problems, and so expands the range of our problem-solving ability.

Conversational Critical Mass

It's challenging to convey the experience of designed serendipity. It's one thing to describe examples, but it's quite another thing to be part of a collaboration where designed serendipity is actually going on. All of a sudden, you feel as though your mind has grown wings. You're liberated from much of the burden of niggling problems, problems that would be routine if only you had access to an expert with just the right skills. It's profoundly enjoyable to instead spend your time concentrating on the problems where you have a special insight and advantage. Designed serendipity is something that must be experienced to be fully understood. But with that said, there is a simple model that can help explain why designed serendipity is important, and how it can qualitatively change the nature of collaboration. That model is a nuclear chain reaction. By reminding ourselves of what happens during a chain reaction we will gain insight into why designed serendipity is important.

The way a chain reaction works is simple. Imagine you have somehow come into possession of a small piece of uranium—uranium-235, the type of uranium that goes into nuclear bombs. (There are several types of uranium, but they don't all undergo nuclear chain reactions. From now on, when I say "uranium" I mean uranium-235.) Uranium atoms, it turns out, aren't very stable. Every once in a while, the nucleus of a uranium atom will disintegrate, spitting out one or more neutrons. That neutron then flies off through the piece of uranium. Uranium, like all solids, only looks solid to the human eye. In fact, at the atomic level it's mostly empty space, and the neutron can travel a long way before it encounters the nucleus of another uranium atom. In a small piece of uranium—say, half a kilogram (about a pound)—the chances are pretty good that the neutron will never encounter another nucleus, and will instead fly all the way out of the piece of uranium, and just keep going. But if the piece of uranium is just a little bit bigger—say, a kilogram—the chances are a fair bit higher that the neutron will smash into the nucleus of another uranium atom. That nucleus then disintegrates, and, it turns out, releases three more neutrons. Now there are four neutrons whizzing through the uranium—it's four because we need to include in our count the original neutron that started the process, which continues to move, even after smashing into the nucleus. Each of those four neutrons is, in turn, likely to smash into four other nuclei, with the result that 16 neutrons are now on the loose. They are likely to crash into still more nuclei, and things rapidly cascade out of control: after 40 collisions like this, we have a trillion trillion neutrons whizzing around. It's because of this incredibly rapid rate of growth that the process is called a chain reaction. Below a certain mass, called the critical mass, a piece of uranium is simply an inert lump of rock. Atoms inside are occasionally decaying and releasing neutrons, but for each such neutron the average number of so-called daughter neutrons caused by further collisions is less than one, and any possible chain reaction quickly dies out. But with just a slightly larger piece of uranium, larger than the critical mass, the average number of daughter neutrons is slightly more than one. And if the average number of daughter neutrons is even a tiny bit larger than one then the chain reaction will take off, and cascade out of control. If the average number of daughter neutrons is 1.1, then after just 200 collisions the uranium will have more than 100 million neutrons flying around inside, causing still more collisions. This is why two apparently similar pieces of uranium will behave in completely different ways. One will lie inert, while another just slightly larger piece will explode with the force of thousands of tons of dynamite. A small increase in size can cause a complete qualitative change in behavior.

Something similar goes on in a good creative collaboration. When we attempt to solve a hard creative problem on our own, most of our ideas go nowhere. But in a good creative collaboration, some of our ideas—ideas we couldn't have taken any further on our own—stimulate other people to come up with daughter ideas of their own. Those, in turn, stimulate other people to come up with still more ideas. And so on. Ideally, we achieve a kind of conversational critical mass, where the collaboration becomes self-stimulating, and we get the mutual benefit of serendipitous connection over and over again. It's that transition that is enabled by designed serendipity, and which is why the experience of designed serendipity feels so different from ordinary collaboration. It occurs when collaboration is scaled up, increasing the number and diversity of participants, and so increasing the chance that one idea will stimulate another new idea. In the Polymath Project, for example, Tim Gowers commented that the main thing that sped up the process was that he and other participants often "found [themselves] having thoughts that [they] would not have had without some chance remark of another contributor." In Kasparov versus the World the same thing happened, with an idea from one team member often sparking ideas from others, enabling the World Team to explore many different directions.

Of course, the chain reaction model shouldn't be taken too literally as a model of collaboration. Ideas aren't neutrons, and the goal of collaboration isn't simply to go "critical," producing a rapidly ballooning number of ideas. We need to, at least occasionally, have the right ideas, ideas that genuinely move us closer to a solution to our problem. It's possible that somewhere in the problem being tackled there's a bottleneck, requiring some key insight that no one in the collaboration is ready to have. Still, the chain reaction model conveys well the qualitative change that takes place when a collaboration "goes critical," when designed serendipity makes the number of ideas being generated in a collaboration jump so high that the process becomes self-sustaining. That jump qualitatively changes how we solve problems, taking us to a new and higher level.

Amplifying Collective Intelligence

Let's take stock of the picture of collective intelligence we're developing. It starts with the idea that within large groups there can be a tremendous amount of expertise, far more than is available from any single individual in the group. Ideally, such groups are extremely cognitively diverse—meaning that they have a wide range of non-overlapping expertise—but their members have enough in common that they can communicate effectively.

Ordinarily, most of this expertise is latent. A good but not great chess player may have individual areas of microexpertise where they equal or surpass the world's best chess players, but in an ordinary chess game that is not sufficient to outweigh the many areas in which they are inferior. But if the group is large enough, and cognitively diverse enough, then the right tools can make it possible for the group to harness such microexpertise when it's needed, and so the group can far exceed the talent of any individual. Designed serendipity can take hold, resulting in a conversational critical mass that rapidly explores a much larger space of ideas than any individual could on their own.

Underlying this broad picture is the fact that collectively we know far more than even the most brilliant individuals. Centuries ago it was, perhaps, possible for a single brilliant individual—an Aristotle or Hypatia or Leonardo—to surpass all others across many areas of knowledge. Today, human knowledge has expanded so that this is no longer possible. Knowledge has been decentralized, and is now held across many minds. Even the most brilliant people, people such as mathematicians Tim Gowers and Terence Tao and chess player Garry Kasparov, have an unsurpassed mastery of only a tiny fraction of our knowledge. Even within their areas of expertise, they're often surpassed in specialized ways by other people, people with particular areas of microexpertise. By restructuring expert attention online tools can enable that microexpertise to be applied when and where it is most needed.

With this point of view in mind, we see that the problem of amplifying collective intelligence is to direct microexpertise where it will be of most use. The purpose of the online tools is to help people figure out where they should direct their attention. The better the tools can direct people's attention, the more successful the collaboration will be. Put another way, the online tools create an *architecture of attention* whose purpose is to help participants find tasks where they have the greatest comparative advantage. Ideally, that architecture of attention will bring the attention of the right expert to the right problems. The more effectively expert attention is allocated in this way, the more effectively problems can be solved. (See the endnotes for discussion of the related idea of the architecture of *participation*, suggested by technology expert Tim O'Reilly.) This view of collective intelligence is summarized in the Summary and Preview box, which also previews many of the ideas about amplifying collective intelligence developed in the remainder of part 1.

Summary and Preview: How to Amplify Collective Intelligence

To amplify collective intelligence, we should scale up collaborations, increasing the cognitive diversity and range of available expertise as much as possible. This broadens the range of problems that can easily be solved. The challenge in scaling up collaboration is that each participant has only a limited amount of attention to devote to the collaboration. That limits the volume of contributions to the collaboration that any one participant can pay attention to. To scale up the collaboration while respecting this limitation, the online tools must establish an architecture of attention that directs

---

Continued on next page

each participant's attention where it is best suited—that is, where they have maximal comparative advantage. Ideally, the collaboration will achieve designed serendipity, so that a problem that seems hard to the person posing it finds its way to a person with just the right microexpertise to easily solve it (or stimulate further progress). Conversational critical mass is achieved, and the collaboration becomes self-stimulating, with new ideas constantly being explored. In the next chapter, chapter 4, we'll see many collaborative patterns that can help achieve these ends, including:

- Modularizing the collaboration, that is, figuring out ways to split up the overall task into smaller subtasks that can be attacked independently or nearly independently. This reduces barriers to entry by new people, and thus broadens the range of available expertise. Modularity is often difficult to achieve, requiring a conscious, relentless commitment on the part of participants.

- Encouraging small contributions, again to reduce barriers to entry, and to broaden the range of available expertise.

- Developing a rich and well structured information commons, so people can build on earlier work. The easier it is to find and reuse earlier work, the faster the information commons will grow.

In chapter 5 we'll examine the limits to collective intelligence. We'll find that for collective intelligence to be successful, participants must be committed to a shared body of methods for reasoning, so disagreements between participants can be resolved, and do not cause permanent rifts. Such a shared body of methods is available in fields such as chess, programming, and science, but not always in other fields. For example, artists may be fundamentally divided over basic aesthetic principles. Such divisions will prevent collaboration from scaling up, and so prevent designed serendipity and conversational critical mass.

How Online Collaboration Goes Beyond Conventional Organizations

Using collective intelligence to solve problems is not new. Historically, groups have used three main ways to solve creative problems: (1) large formal organizations, such as the hundreds or thousands of people who may be involved in creating a movie, say, or a new electronic gadget; (2) the market system; and (3) conversation in small informal groups. In the remainder of this chapter we'll investigate how online tools can take us beyond these three existing ways of doing group problem solving.

To understand how online collaboration goes beyond conventional organizations, consider a movie production. A modern blockbuster movie may employ hundreds or even thousands of people—the 2009 movie *Avatar* employed 2,000 people. But unlike the participants in Kasparov versus the World or the Polymath Project, each employee has their own assigned role in the production. An employee in the movie's art department won't usually give advice to a violin player in the orchestra. Yet that's exactly the kind of decision making that happened in Kasparov versus the World. Recall the critical move number 26 suggested by Yasha. In movie terms, it was as though an unknown stranger had wandered on set, made a crucial suggestion to the director, completely changing the course of the movie, and then wandered off.

Of course, there are such stories in the movies. Actor Mel Gibson got his big break when a friend who was auditioning for the movie *Mad Max* asked to be driven to the audition. Gibson wasn't auditioning, but had gotten into a brawl at a party the night before, and had bruises all over his face. The casting agent decided that was the look the movie needed, and Gibson was invited back, completely changing the movie, and launching him on the path to international stardom.

In the world of movies this is an unusual story. But in Kasparov versus the World this kind of occurrence wasn't a lucky one-off, it was the essence of the way the World Team played. There was no preplanned, static division of labor, as in a conventional organization. Instead, there was a dynamic division of labor, in which every player on the World Team had the opportunity, at least in principle, to be involved in every move.

Let me make more precise what I mean by a dynamic division of labor. It's a division of labor where all participants in a collaboration can respond to the problems at hand, as they arise. Zacary Brown saw ASSET's problem, and realized he could solve it. Yasha followed along the World Team's progress, and realized he had a special insight at move 26. And all participants in the Polymath Project could follow the rapidly evolving conversation, and jump in whenever they had a special insight. In conventional offline organizations, such flexible responses are usually only possible in small groups, if at all. In larger groups different group members focus on their own preassigned areas of responsibility. Online tools change this, making it possible for large groups to harness each participant's special areas of microexpertise, just-in-time as the need for that expertise arises. That's what I mean by a dynamic division of labor. Ideally, as we saw earlier, this will lead to designed serendipity. But even when that doesn't happen, the dynamic division of labor is still strikingly different from the conventional static division of labor.

None of this is to deny the value of a static division of labor. We've achieved enormous improvements in our ability to manufacture goods by improving the static division of labor—think of Henry Ford's assembly line, or even Adam Smith's hypothetical pin factory. But while such a division of labor is well suited to the manufacture of goods, using a predictable and repetitive process, it's been less useful in solving hard creative problems. The reason is that in creative work it's often the unplanned and unexpected insights and connections that matter the most. In many cases, what makes a creative insight important is precisely the fact that it combines ideas that previously were thought to be unrelated. The more unrelated, the more important the connection—recall the astounding connection Einstein and Grossmann made between gravity and Riemannian geometry. Because of this, the greatest creative work can't be planned as part of a conventional static division of labor. No one could have predicted that Kasparov versus the World would unfold the way it did, and so it wasn't possible to anticipate that Krush's special microexpertise would be needed to cope with the position that occurred at move 10. And it certainly wasn't possible to anticipate the need for Yasha on move 26. It was only possible to do that division of labor dynamically, as the situation arose.

The reason this all matters is that for hard creative problems, until recently we've had to rely on the genius of individuals and small groups, and lucky occasional serendipitous interactions. This limits the range of expertise that can be brought to bear. Even in a task such as movie making, with its reputation for being free-wheeling, the major creative decisions are mostly made by a small number of people. Now, it should be said that modern organizations aren't completely wed to the static assembly-line style of doing things. They often achieve a dynamic division of labor on a small scale, with small groups working in creative teams. That happens, for instance, in movie productions, and it also happens in many other creative organizations, including celebrated organizations such as Lockheed Martin's Skunk Works, or the Manhattan Project, which developed the atomic bomb. Management techniques such as Total Quality Management and lean manufacturing incorporate ideas that help enable a more dynamic division of labor—a famous example is the way Toyota delegates to factory workers great responsibility for finding and fixing manufacturing defects on the fly. What is new about online tools is that they make it far easier to do such a dynamic division of labor on a large scale, bringing the expertise of much larger groups to bear on hard creative problems.

The distinction between dynamic and static division of labor also illuminates the difference between online collaborations and conventional large-scale scientific collaboration. Consider, for example, the collaboration of 138 particle physicists whose work led to the 1983 discovery of the Z boson, a new fundamental particle of nature, at Europe's CERN particle accelerator. Unlike Kasparov versus the World or the Polymath Project, each of the people in the CERN collaboration was hired to fill a set role. The roles ranged over many carefully chosen specialties, from engineers whose job was to cool down the particle beam, to statisticians whose job was to make sense of the complex experimental results. Such specialized collaborations can accomplish remarkable things, but with their relatively fixed roles and static division of labor they leave a great deal of microexpertise latent, and show little flexibility in their purpose. Their inflexibility means that while they can do extremely important science, it's not a model that can easily be adapted to the more fluid ends characteristic of much of the most creative scientific work.

How Online Collaboration Goes Beyond the Market

One of humanity's most powerful tools for amplifying collective intelligence is the market system, and we can learn much about online collaboration by comparing it to the market. Of course, the market is so familiar that it's tempting to take it for granted, and to focus only on examples where it amplifies collective stupidity, such as the crashes of 2008 and 1929. But most of the time the market really does amplify our collective intelligence. In his book *The Company of Strangers*, the British economist Paul Seabright tells how two years after the breakup of the Soviet Union he met with a senior Russian official who was visiting the UK to learn about the free market. "Please understand that we are keen to move towards a market system," the Russian official said, "But we need to understand the fundamental details of how such a system works. Tell me, for example: who is in charge of the supply of bread to the population of London?"

The familiar but still astonishing answer to this question is that in a market economy, everyone is in charge. As the market price of bread goes up and down, it informs our collective behavior: whether to plant a new wheat field, or leave it fallow; whether to open that new bakery you've been thinking about opening on the corner; or simply whether to buy two or three loaves of bread this week. The prices are signals to help coordinate the actions of suppliers and consumers: as demand for a good goes up, so does the price, motivating new suppliers to enter the market. The result is a marvelous dance of actions that puts food on our tables, cars in our garages, and smartphones in our pockets. Familiarity makes us take this for granted, but the dance is really a miraculous mass collaboration, mediated so smoothly by the market that it's only noticed when absent.

What makes prices useful is that, as emphasized by the economist Friedrich von Hayek, they aggregate an enormous amount of hidden knowledge—knowledge that would otherwise not be apparent to all the people interested in the production or consumption of goods. By using prices to aggregate this knowledge and inform further actions, the market produces outcomes superior to even the brightest and best informed individuals. It enables a dynamic division of labor: if flooding wipes out the wheat crop in much of the United States, then the price will rise, and other suppliers of wheat will respond by working hard to increase the supply.

Markets and the price system thus have many of the properties we've identified in online collaboration. In contrast to conventional offline organizations, they use both a dynamic division of labor and designed serendipity. But online collaborations such as the Polymath Project go beyond offline markets in the complexity of the problems under consideration, and in the speed with which unanticipated problems may be posed and addressed. Even if you have no interest in mathematics, it's easy to appreciate the rich flavor of this "dumb question" posed by Polymath participant Ryan O'Donnell:

> Can someone help me with this dumb question?
> Suppose $A = B$ are the family of sets not including the last element n. Then A and B have density about $1/2$ within $K N_{n,n/2-k/2}$. (We're thinking $k(n) \rightarrow \infty, k(n)/n \rightarrow 0$ here, right?) [...]

That's just the beginning of the question; it's a far cry from "What's the price of bread?" O'Donnell's question is far too specialized and context-dependent to be addressed by a conventional offline market. He could, perhaps, have taken out an advertisement in a mathematics journal asking for help, but the bother would have been greater than the benefit. In an online collaboration such as the Polymath Project such a question can occur to someone, be broadcast to other participants, and answered, all within minutes or hours. Online tools thus combine the dynamic division of labor and designed serendipity found in markets with the flexibility and spontaneity of everyday conversation. This combination makes them a big step forward from offline markets, and, in particular, makes them well suited to attacking hard creative problems.

So far I've focused on conventional offline markets. Of course, in recent years markets have adopted the internet and other modern communications technologies, and as they've done so they've changed and become more complex. Increasingly, they too can be used to address very specialized and context-dependent questions. In this sense online tools are gradually subsuming and extending markets. Something similar is also going on in the conventional organizations we discussed in the last section: online tools are increasingly used as the command and control infrastructure in those organizations. And so online tools can subsume and extend both conventional markets and conventional organizations. And, as we'll see shortly, they can also subsume and extend the third historical form of collaboration, small group conversation. In each case, the online tools are enabling architectures of attention that go beyond what is possible in offline methods of collaboration.

How Online Collaboration Compares to Offline Small-Group Conversation

In many respects online collaborations such as the Polymath Project and Kasparov versus the World resemble offline small-group conversation. As we'll see, in some ways offline conversation is actually genuinely better than online collaboration, while in other ways, it is distinctly inferior. But before we compare the two, let's first clear the air by disposing of two common but fallacious arguments that purport to relate online collaboration to offline conversation.

The first fallacy is to think that online collaboration is somehow similar to dreary committee work. Sometimes people hear about a project such as the Polymath Project, and their mind leaps to the unflattering stereotypes we associate with committees—"A camel is a horse designed by committee," and so on. It's true that many committees squelch creativity and commitment. But it doesn't follow that online collaboration has the same problems. When you look closely at projects such as the Polymath Project and Kasparov versus the World, they don't seem much like dysfunctional committees. Instead, they are vibrant communities filled with creativity and commitment.

How do such collaborations escape the problems of dysfunctional committees? Understanding why some groups work well while other don't is a complex problem, and I won't comprehensively address this question here. But there are two powerful factors that help explain why online collaboration often works well where a committee would not. First, committees are often made up of people who've been dragooned to sit on them, while collaborations such as the Polymath Project are filled with enthusiastic volunteers. That passionate commitment makes a big difference. Second, while a committee can be greatly slowed down by a few obstructive members, online collaborations can often ignore those people. In the Polymath Project, for example, it was easy for well-informed participants to ignore the occasional well-intentioned but unhelpful contribution. Collaborating online is simply not the same as committee work.

A second fallacy sometimes put forward by skeptics of online collaboration is that it's always possible to replace online collaborations by equivalent offline collaborations. For example, they might argue that given enough patience and a room full of mathematicians, you could do an offline "simulation" of the Polymath Project. There are two problems with this argument. The first is that, as a practical matter, it's far easier to get together online than offline. So the objection is a little like saying that the invention of the automobile or passenger train changed nothing about travel, because people had always been able in principle to use a horse and buggy to travel long distances. The observation is true, but has little practical importance for how people actually behave. The second problem is that human behavior in a room full of mathematicians would in practice be dramatically different than in the Polymath Project. To pick one of many examples of differences: offline, if someone speaks with you when you're tired and cranky, you may not understand what they said; online, you can read and reread at your leisure, when you're alert and enthused. Because of these and many other differences, you can only do the offline simulation if you make unrealistic assumptions about how humans would behave in the room. This is not to say a room full of mathematicians couldn't collaborate to do remarkable work. But it wouldn't be using a Polymath-style process, it would be using a different architecture of attention. Online tools really do enable us to collaborate in new ways.

With those two fallacies out of the way, what of the ways in which offline conversation is genuinely superior to online collaboration? One especially stands out: the rich nature of face-to-face contact.

Body language, facial expression, tone of voice, and regular informal contact are all tremendously important to effective collaboration, and cannot be replaced. With people you like, in-person conversation is enjoyable and stimulating, and online collaboration loses something by contrast. Of course, this loss is gradually being offset by more expressive collaborative technologies — a tool such as Skype video chat is remarkably effective as a way to collaborate. Over the longer run ideas such as virtual worlds and augmented reality may even make online contact better than face-to-face contact. Still, today the online experience of direct person-to-person collaboration lacks much of the richness of offline collaboration. It's tempting to conclude that online collaboration can't be as good as offline.

The trouble with this conclusion is that it ignores the problem of how you find the right person to work with in the first place. This is perhaps because finding that right person has historically been such a hard problem that we usually don't bother. Offline, it can take months to track down a new collaborator with expertise that complements your own in just the right way. But that changes when you can ask a question in an online forum and get a response ten minutes later from one of the world's leading experts on the topic you asked about. In creative problem solving, it's often better to have a terse twenty-minute text-only interaction with an expert who can solve your problem with ease, rather than weeks of enjoyable face-to-face discussion with someone whose knowledge is not much different than your own. And, in any case, you don't have to make this choice. In practice, you can use relatively impersonal tools to find the right person or people for the problem at hand, and more expressive tools such as video chat, virtual worlds, and augmented reality to make working with that person or people as effective as possible.

To put it another way, the big advantages of online collaboration over offline conversation are in scale and cognitive diversity. Imagine that the people at ASSET India had gotten together a group to brainstorm ideas for wireless routers. Unless they were extremely lucky, the group would not have contained anyone with the same kind of expertise as Zacary Brown. By increasing the scale of collaboration, online tools expand the range of available expertise, reducing the chance that the group will be blocked by a problem that no one in the group can solve. Ideally, designed serendipity and conversational critical mass will occur, enabling the group to explore in depth a far wider range of ideas than is possible in a small group, with its limited expertise.

How do online tools enable conversation to be scaled up? The obvious answer is that online tools make it easier for experts around the world to get together as part of a group. That is important, but it's only a small part of what's going on. In fact, by using a carefully designed architecture of attention, online tools enable collaborations to involve far more people than is practical in offline conversation. Let me describe how this worked in the Polymath Project. Superficially, the format of the Polymath Project, based on comments on blogs, seems similar to discussions of mathematics in face-to-face conversation. But it goes further in three important ways. First, when working online people pre-filter their comments more than in ordinary mathematical conversation. In offline conversations even the best mathematicians have long pauses, need to backtrack, and occasionally get confused. In the Polymath Project most comments distilled one point in a relatively sharp way. Second, as a reader it's easy to skip rapidly over blog comments. When you're face to face, if you don't understand what someone's saying, you may be stuck listening to them speak incomprehensibly for ten minutes. But on a blog you can glance at a comment for a few seconds, take note of the general idea, and move on. Third, when you skip a comment you always know that you can return to it later. It's archived, and easily findable using search engines. The overall effect of these three differences is to scale up the number of people who can participate in the conversation. By increasing the scale of conversation the blog medium gives us access to the best ideas from a more cognitively diverse set of participants, and so designed serendipity and conversational critical mass are more likely to occur.

There is, however, an inherent trade-off in scaling up collaboration. On the one hand, a collaboration should involve the largest and most cognitively diverse group of participants possible. On the other hand, once the collaboration gets large enough participants cannot possibly pay attention to everything that's going on. Instead, they perforce *must* begin paying attention to only some of the contributions. Ideally, the architecture of attention will direct participants to places where their particular talents are best suited to take the next step—where they have maximal comparative advantage. So each participant sees only part of the larger collaboration. As a simple example, InnoCentive classifies Challenges into subject areas, to help participants find the Challenges of most interest to them. In the next chapter, we'll see some more sophisticated ways of helping people decide where to direct their attention. In this way, it's possible to scale beyond the point where each participant must pay attention to the entire collaboration. Put another way, the art of scaling is to filter contributions so each participant sees only the contributions they personally will find most valuable and stimulating; the important thing isn't what we see, it's what we get to ignore. The better the filters, the better our attention is matched to opportunities to contribute. In a nutshell, an ideal architecture of attention enables the largest, most cognitively diverse group to best utilize the limited available attention so that at any given time each participant is maximizing their comparative advantage. Collaborations such as the Polymath Project go only a small part of the way toward this goal. By using a better architecture of attention it is possible to scale collaboration even further than the Polymath Project. In the next chapter we examine several patterns that can be used to scale up online collaborations, and to make better use of the available expertise.

CHAPTER 4

Patterns of Online Collaboration

On August 26, 1991, at 2:12 am, a 21-year-old Finnish programming student named Linus Torvalds posted a short note to an online forum for programmers. It read, in part:

> I'm doing a (free) operating system (just a hobby, won't be big and professional like gnu) for 386(486) AT clones ... I'd like to know what features most people would want. Any suggestions are welcome, but I won't promise I'll implement them :-)

Just 14 minutes later, another user responded with the words "Tell us more!" and asked several questions. Nearly six weeks later, on October 5, Torvalds posted a second note, announcing that the code for his operating system—soon to be dubbed Linux—was now publicly available. He wrote in the announcement:

> This is a program for hackers by a hacker. I've enjouyed doing it, and somebody might enjoy looking at it and even modifying it for their own needs. It is still small enough to understand, use and modify, and I'm looking forward to any comments you might have.

Torvalds was an unknown, a student working in relative isolation at the University of Helsinki, not part of some hip Silicon Valley startup company. Still, what he'd announced was interesting to many hackers. The operating system is the nerve center of a computer, the piece that makes the rest of it tick. Handing a hardcore hacker the code for an operating system is like giving an artist the keys to the Sistine Chapel and asking them to redecorate. Shortly after Torvalds's post, a Linux activists mailing list was formed, and just three months later the mailing list had grown to 196 members.

Torvalds not only made the code for his operating system freely available, he also encouraged other programmers to email him code for possible incorporation into Linux. By doing this, Torvalds initiated the formation and rapid growth of a community of Linux developers—programmers who collectively helped him improve Linux. By March of 1994, 80 people were named as contributors in the Linux Credits file, and people were contributing code at an astronomical rate. In 1995, the company Red Hat formed, marketing one of the first commercially successful versions of Linux; in 1999, Red Hat went public on the New York Stock Exchange, with a market value of 3 billion dollars by the end of its first day of trading. By early 2008, the Linux kernel—the core part of the Linux operating system—contained nearly 9 million lines of code, written by a collaboration of more than 1,000 people. It is one of the most complex engineering artifacts ever constructed.

Linux has become so widespread that it's easy to take it for granted. Although Microsoft Windows remains the dominant operating system for home and office use, in many other areas Linux surpasses it. Companies such as Google, Yahoo!, and Amazon all use enormous Linux clusters, containing tens or hundreds of thousands of computers. In Hollywood animation and visual effects companies, Linux is the dominant operating system, surpassing Windows and MacOS and playing a major role at Pixar, Dreamworks, and Industrial Light and Magic. In the consumer electronics industry, companies such as Sony, Nokia, and Motorola use Linux in everything from mobile phones to televisions. This ubiquity makes it easy to forget how remarkable the story of Linux is. Imagine that in 1991 a 21-year-old Finnish programming student had approached you, telling you that he'd written the core of an operating system and was planning to release the code, and oh, by the way, he was hoping to recruit a volunteer army of programmers to improve it. You'd think it was ludicrous. It was ludicrous. So ludicrous that not even Torvalds himself imagined it would happen.

Linux is an example of open source software. Open source software projects have two key attributes. First, the code is made publicly available, so anyone can experiment with and modify the code, not just the original programmer. Second, other people are encouraged to contribute improvements to the code. This might mean sending in a bug report when something goes wrong, or perhaps suggesting a change to a single line of code, or even writing a major code module containing thousands of lines of code. The most successful open source projects recruit large numbers of contributors, who together can develop software far more complex than any individual programmer could develop on their own. To give you some idea of the scale, in 2007 and 2008 Linux developers added an average of 4,300 lines of code per *day* to the Linux kernel, deleted 1,800 lines, and modified 1,500 lines. That's an astounding rate of change—on a large software project, an experienced developer will typically write a few thousand lines of code per *year*.

Of course, most open source projects have fewer contributors than Linux. A popular repository of open source projects called SourceForge houses more than 230,000 open source projects. Nearly all those projects have only one or a few contributors. But a small number of projects have captured the imagination of programmers, drawing in tens, hundreds, or thousands of contributors.

Open source started in the programming world, but it isn't fundamentally about programming. Rather, open source is a general design methodology that can be applied to any project involving digital information. If you're an architect, for example, you can do open source architecture: simply share the designs for your buildings freely, and encourage others to contribute improvements. In 2006, an architect named Cameron Sinclair and a journalist named Kate Stohr launched the Open Architecture Network, which is creating an online community for open source architecture—a kind of Source-Forge for architecture. As of early 2010, the site contained more than 4,000 projects, many with floor plans, discussions of building materials, photographs of finished buildings, and so on, all available for reuse and improvement by others. The site focuses especially on designs for use in the developing world, and Sinclair and Stohr hope that it will help the best architectural ideas and innovations spread more quickly. An example is shown in figure 4.1, the design for a primary school built in Gando, a town of 3,000 people in the tiny country of Burkina Faso (previously known as Upper Volta) in

West elevation   East elevation

North elevation

South elevation

Figure 4.1. Top: a primary school in the town of Gando, in the country of Burkina Faso in West Africa. Bottom: one of several design documents for the school, freely available for download from the Open Architecture Network. Other people may use the design documents, and modify them for their own needs. Credit: Siméon Duchaud / Aga Khan Award for Architecture.

West Africa. The design comes complete with floor plans, elevations, and many other design details, as well as photos of the finished school.

It's not just architecture that can be open source. If you're a digital artist, you can do open source art: share the files for your digital art freely, and encourage others to contribute improvements. If you're a biologist you can do open source biology: share DNA designs for living things, and encourage others to contribute improvements. There's a community of biologists doing exactly that. If you're writing an encyclopedia, you can share the text of your encyclopedia articles freely, and encourage others to contribute improvements. That's how Wikipedia is written: Wikipedia is an open source project. The underlying pattern in all these projects is the same: share your digital design, and encourage other people to contribute changes. The Polymath Project doesn't quite follow this pattern, but it does use similar ideas, creating an online space where people can share their ideas, and work to improve other people's ideas.

So far in this book, we've looked at several examples that show how online tools can make groups smarter. Open source collaborations usually have different purposes: they're about giving people the freedom to improve and modify other people's work, and—for big projects, such as Linux and Wikipedia—about enabling groups to create projects more complex than any individual could create on their own. This difference in purpose is reflected in the fact that while Wikipedia is impressive, for many subjects the world's top experts could write better articles. Similarly, the code for Linux merely needs to be good enough to work, it doesn't need to be of the highest quality throughout. But despite this difference in intent from our earlier examples, open source can still teach us much about how to amplify collective intelligence. In particular, open source collaborations have been superbly effective at scaling up, and so increasing the cognitive diversity and range of microexpertise available to the collaboration. In this chapter we'll identify four powerful patterns that open source collaborations have used to scale. (1) a relentless commitment to working in a modular way, finding clever ways of splitting up the overall task into smaller subtasks; (2) encouraging small contributions, to reduce barriers to entry; (3) allowing easy reuse of earlier work by other people; and (4) using signaling mechanisms such as scores to help people decide where to direct their attention. These patterns can be incorporated into any architecture of attention, and so be used to amplify collective intelligence.

The Importance of Being Modular

To understand how open source collaborations scale, let's look at a time when the Linux collaboration almost failed to scale, a time when the Linux developer community almost fractured into two separate camps, working on two separate versions of Linux. The incident started innocuously, on September 29, 1998, with a post to the Linux kernel mailing list by developer Michael Harnois. Harnois wrote to say that he was having problems with part of the Linux display system. This kind of complaint was not unusual—indeed, such complaints are the grist that Linux developers use to improve the code—and a well-respected Linux developer named Geert Uytterhoeven quickly replied to Harnois. Uytterhoeven told him not to waste his time, that the problem had already been fixed, and the only reason Harnois was having problems was because the code fixing the problem wasn't yet included in the official Linux code base, maintained by Linus Torvalds.

So far, this was business as usual. But what Uytterhoeven added next sparked a major blowup. He told Harnois that while the fix for his problem wasn't yet in the official code base, he could get a copy of the fix from a website called VGER. VGER was a service started as a mirror (that is, a copy) of the official Linux code, an alternate location where people could download Linux, in case the main site was down or hard to reach. But some Linux developers were growing unhappy with Torvalds, believing that he wasn't integrating their contributions fast enough into the official Linux code base. The group of volunteers running VGER, on the other hand, were accepting some of those contributions, and it was quietly known that the "VGER Linux" was starting to run ahead of the official Linux in crucial ways.

Less than two hours after Uytterhoeven's post, Linus Torvalds replied with a terse post to the mailing list, saying Harnois was "not wasting time," and that VGER was irrelevant to Linux development. Torvalds's post touched off an avalanche of responses, with some of the most respected Linux contributors complaining loudly that this was not the first time he had failed to integrate an important contribution into the official Linux code. Several complained that they had sent Torvalds code contributions multiple times without receiving any acknowledgment, sometimes even for work they'd done at his request. Torvalds, for his part, also expressed frustration:

> Quite frankly, this particular discussion (and others before it) has just made me irritable, and is ADDING pressure. Instead, I'd suggest that if you have a complaint about how I handle [contributions], you think about what I end up having to deal with for five minutes.

> Go away, people. . . . I'm not interested, I'm taking a vacation, and I don't want to hear about it any more. In short, get the hell out of my mailbox.

To be successful, a collaboration must divide the problem it's attacking into tasks that can be done by single individuals. By the time of this blowup the Linux community had grown so much that the task of reviewing and integrating code submissions was beyond Torvalds (or probably any single person). In the words of one of the Linux developers involved in the imbroglio, Larry McVoy, "Linus doesn't scale." As a result, the Linux development community was no longer working effectively, and was in danger of fragmenting into two or more separate communities. This wasn't because Torvalds or anyone else was doing anything wrong. Instead, it was a consequence of success: the community had grown so much that the old way of doing things no longer worked.

The obvious way to solve the problem was to split the task of approving code contributions between several people. But some Linux developers worried that Torvalds's broad understanding of the Linux kernel might be essential to reviewing and approving code contributions. Might allowing others to approve contributions actually damage Linux? Perhaps some essential but previously tacit functionality in the Linux collaboration might be lost. Fortunately, those fears were not borne out. After a heated online discussion, and a face-to-face meeting of some of the leading Linux developers, including Torvalds and the creator of VGER, Dave Miller, Torvalds agreed to delegate more decision making to lieutenants, and this went ahead without any evident ill effects.

In some collaborations it's easy to divide the problem being attacked into smaller tasks. Recall the galaxy classification project Galaxy Zoo, which we met in the opening chapter. Galaxy Zoo asks contributors to answer questions about just one galaxy at a time, dividing the problem of classifying galaxies up into millions of tiny tasks. That's a simple but effective way of dividing Galaxy Zoo's overall problem.

Sometimes, however, this kind of modularity is much harder to achieve. In the Polymath Project, work was carried out through comments on blog postings. In the early days of the project, it was easy for interested mathematicians to join the discussion. But the number of comments quickly climbed, eventually reaching 800 comments and 170,000 words. For outsiders this was a daunting barrier to entry, since the comments weren't organized in a way that would allow them to jump into the discussion without first understanding the bulk of the earlier contributions. Although the Polymath Project was a large collaboration by conventional standards in mathematics, with contributions from 27 people, it would likely have been even larger had the discussion been less monolithic and more modular. That, in turn, would have increased cognitive diversity, making a greater range of expertise available to the collaboration.

Is this monolithic narrative style an inevitable feature of collaborations such as the Polymath Project? Or is it possible to devise a more modular approach that breaks the collaboration up into subprojects? We can get insight into these questions by taking a closer look at large open source projects such as Linux. Those projects have not achieved modularity easily or by chance, but by working very, very hard at it. They've made a conscious commitment to be modular, and then relentlessly followed through on that commitment, even when it required a great deal of work. We've seen an example of this in the way the Linux community responded to the VGER crisis. But even more impressive, albeit in a quieter way, is the day-in, day-out commitment the Linux developer community shows to being modular. As an example, the original code base for the Linux kernel didn't have the sort of simple modular structure that would make it easy for potential developers to get involved in improving the code. For Linux release 2.0 the entire Linux code base was substantially rewritten and reorganized to make it modular. That perhaps sounds easy on paper, but it required a huge coordinated effort by the Linux developers. Here's how Torvalds explained it:

> With the Linux kernel it became clear very quickly that we want to have a system which is as modular as possible. The open-source development model really requires this, because otherwise you can't easily have people working in parallel.
>
>   . . .
>
> With the 2.0 kernel Linux really grew up a lot. This was the point that we added loadable kernel modules. This obviously improved modularity by making an explicit structure for writing modules. Programmers could work on different modules without risk of interference. I could keep control over what was written into the kernel proper. So once again managing people and managing code led to the same design decision. To keep the number of people working on Linux coordinated, we needed something like kernel modules. But from a design point of view, it was also the right thing to do.

This pattern of conscious, relentless modularity is seen in most large open source collaborations. It's often required even in projects where modularity looks as though it would be easy to achieve, such as Wikipedia. On the surface, Wikipedia appears to be merely a collection of encyclopedia articles, with a simple, natural modular structure: the writing is naturally divided up between the different articles. But that superficial modularity is only part of the story. Writing an encyclopedia involves many tasks beyond editing the articles, and that additional complexity is reflected in Wikipedia's structure. Perhaps the simplest example is that every Wikipedia article has an associated "Talk" page. If you don't know what a Wikipedia Talk page is, start up your web browser, and load Wikipedia's "Geology" article (http://en.wikipedia.org/wiki/Geology). At the top of the page, you'll notice a tab labeled "Discussion." Click on the tab, and you'll be taken to the Talk page for the "Geology" article. That's where discussion *about* the article goes on among Wikipedia editors: discussion of shortcomings in the article, discussion of how the article can be improved, and even discussion of whether the article should exist in the first place. Such Talk pages are a locus for conversations about many tasks that are essential for Wikipedia to work properly, but that can't be carried out on the article pages. Beyond the Talk pages, Wikipedia also has a vast array of other special pages, each aimed at specific tasks. The "Village Pump" page, for example, is for discussion of Wikipedia policy, technical issues, and so on. There's a page listing articles being considered for deletion from Wikipedia. Many Wikipedia pages deal with topics only of interest to the Wikipedia community itself. Some of these pages are funny: there's a 1,181-question test to see if you're a Wikipediholic (for anyone who willingly sits through the entire test, I think the answer is obviously "yes"); a list of articles with freaky titles ("22.86 Centimetre Nails," the metric version of the band "Nine Inch Nails," now unfortunately deleted); and many others. Some of the pages are sad: there is a page listing deceased Wikipedians, with links to their user pages, where you will often find grieving communities of friends and family. Wikipedia is not an encyclopedia. It's a virtual city, a city whose main export to the world is its encyclopedia articles, but with an internal life of its own. All those pages—the Talk pages, the special pages, the community pages, and the articles themselves—reflect vital tasks within Wikipedia, and help break up the enormous problem of running an encyclopedia into many smaller tasks. And, as in a well-run city, this division wasn't determined in advance by some central committee, but rather sprang into existence organically, in response to the needs and wants of Wikipedia's "residents"; the editors who write Wikipedia.

When this pattern of conscious, relentless modularity isn't used, open source collaboration doesn't scale. There have, for example, been many failed attempts to use wikis and an open source approach to write a good quality novel. One high-profile attempt was the Million Penguins project, run by the book publisher Penguin in February and March of 2007. The idea was to recruit writers to produce a collaborative novel using wiki software. Judged by the number of people who contributed (1,500), the project was a success. But those people never managed to work effectively together, and as a work of literature the result was a failure. Early on in the project, one of the coordinators, Jon Elek, wrote, "I'll be happy so long as it manages to avoid becoming some sort of robotic-zombie-assassins-against-African-ninjas-in-space-narrated-by-a-Papal-Tiara." The actual novel was far stranger. Here's a short sample to give you the flavor:

> There was no possibility of taking a walk that day ... a swim, perhaps, but not a walk—for Artie was a whale, a humpback whale, to be precise, at least in these moments. It was a sunny day, and Artie would have worn his sunglasses, but being a whale meant he didn't have ears, which made it difficult for his sunglasses to stay on. No matter, he thought, at least he was young and strong.

It's easy to see why Penguin carried out this experiment. Wikis have been successfully used to produce not just an encyclopedia, but also many other reference works, from the fabulous Muppet Wiki (muppet.wikia.com) to the US Intelligence Community's Intellipedia (no publicly accessible URL for that one, sorry!). Superficially, a novel looks quite similar to an encyclopedia or other reference work. But the degree of modularity sufficient to produce an encyclopedia is not sufficient to write a first-rate novel, because it leaves some essential tasks unperformed. Every sentence in a novel has a potential relationship to every other sentence, a potential relationship to each story arc within the novel, and a relationship to the overall story arc. A good author is aware of all these relationships, and uses them to achieve resonance and reinforcement between different parts of the story, and to avoid dissonance and incoherence. To write a good novel, one of the tasks always before you is to compare the sentence you're writing *right now* to all these other parts of the novel, thinking about whether it enhances or detracts from the whole of the novel. For collaborative writing to succeed, someone must keep track of all these possible relationships. Yet wikis don't provide any natural way of breaking down the problem of keeping track of these relationships. So while wikis may work well for short, independent articles such as appear in a reference work, they don't work as a collaborative medium for longer pieces of writing. Still, collaborative technology is in its early days. My bet is that one day soon a technology for online collaboration will be developed, probably not too dissimilar from a wiki, but making it easy to keep track of relationships between different parts of a novel. That will be a big step toward the first good novel written by an open source collaboration. (Of course, managing these relationships is only part of the challenge; in the next chapter we'll meet more difficulties.)

We've seen how Wikipedia and similar reference wikis use a carefully chosen page structure to modularize. Another approach to modularity is illustrated by the way work on the Firefox web browser is organized. If you're not familiar with Firefox, it's a popular alternative to the Internet Explorer web browser. Like Linux, Firefox is an open source project. But the Firefox developers organize their work using a different approach from that of both Linux and Wikipedia. In particular, they organize much of their work using a tool known as an *issue tracker*. To understand how the issue tracker works, imagine you're a user of Firefox who's run into a bug. For example, a bug I've sometimes noticed is this: in my list of Firefox bookmarks, the little pictures (called favicons) alongside my bookmarks sometimes get mixed up. That is, the wrong picture will show up beside a bookmark, or seemingly random favicons from other sites will show up for no apparent reason. I've no idea why this happens, and it's only a minor irritation in an excellent product, but it can be a little confusing. Anyway, having noticed this bug, you decide to help the Firefox project out by reporting it. To do this, you visit Firefox's online issue tracker, a website where you can enter a description of the problem you're having, and any other details that might come in handy to people trying to fix the bug: what webpage you were browsing when you noticed the bug, what operating system you use, what version of Firefox, and so on.

I asked you to imagine doing this, but actually you don't need to imagine it. I checked the Firefox issue tracker, and someone going by the name Bob did exactly what I've just described on January 11, 2008. Once he submitted his report for the favicon bug, it quickly made its way to the issue tracker's list of "Hot Bugs." The Hot Bug list is Firefox Central Station, with many of the developers who work on Firefox watching the list closely. When they see a bug they think they can help fix, they jump in. For Bob's favicon bug a discussion thread quickly started. Reading through the discussion, you learn that the bug is surprisingly subtle, and actually involves more than one problem in the Firefox code. Dozens of people eventually got involved before the bug was conclusively fixed.

The issue tracker isn't just for fixing bugs, it's also used to propose and implement new features. If you want to suggest a new feature in Firefox, you can go to the issue tracker, suggest the feature, and a conversation will begin. If enough people want the feature, someone will start to code it up. The issue tracker thus acts as a smorgasbord of problems and ideas, each with their own attached conversational threads. It's a great way of modularizing work: by organizing participants' attention around single issues, the issue tracker limits the scope of conversation, and so limits the amount of attention people must invest to participate. Instead of having to understand the entire previous discussion, as in the Polymath Project, participants just need to understand the issue at hand. This enables many more people to get involved, and for the collaboration to benefit from a much broader range of expertise. In other words, the payoff from relentless and conscious modularity is that no one needs to understand the whole project in detail, but can instead contribute where they are best able. The overall effect is like a virtual shipyard. Many different people are spread all over the place, contributing to the different parts of the ship, in separate efforts, each modest in size and scope. But the aggregate product is remarkable.

Of course, modularity isn't the end of the story. It's merely a single pattern that helps scale up collaboration. The modular units are the atoms of attention out of which the architecture of attention is built. The ideal, as we've seen, is to create an architecture where those modular units are arranged in such a way that each participant sees those tasks where they have greatest comparative advantage, and so can make the greatest contribution. Existing tools, such as blogs, wikis, and issue trackers do this only imperfectly. But over the long run we'll gradually see the emergence of a design science of attention, which helps us build tools that best use the available expert attention.

And what of Linux? Linus Torvalds long ago gave up trying to follow the entire Linux kernel developer community. In May 2000, a poster to the Linux kernel mailing list complained that Torvalds wasn't replying to his posts. Torvalds replied as follows:

> Note that nobody reads every post in linux-kernel. In fact, nobody who expects to have time left over to actually do any real kernel work will read even half. Except Alan Cox [one of Torvalds's lieutenants], but he's actually not human, but about a thousand gnomes working in under-ground caves in Swansea. None of the individual gnomes read all the postings either, they just work together really well.
>
> Anyway, some of us can't even read all our personal email, simply because we get too much. I do my best.

Linux has grown greatly since Torvalds wrote that post. Today no one, not even the superhuman Alan Cox, can follow all the work going on. The beauty of the Linux collaboration is that it's organized so no one needs to.

Radical Reuse and the Information Commons

Modularity is important, but there's an even more basic pattern of collaboration underlying open source: the ability of open source programmers to reuse and modify one another's work. This may seem so obvious as to be unworthy of consideration, but it has some surprising consequences. The obvious impact, of course, is that programmers don't have to start from scratch, but instead can build on and incrementally improve what others have done. Effectively, open source programmers are building a publicly shared information commons. This commons isn't located anywhere in particular, but rather consists of all the open source code distributed in myriad locations across the internet. This enables a dynamic division of labor, in which code from one person can later be improved by other people whom they have never met, with expertise and needs they may never even have heard of. The richer the information commons becomes, the more powerful a foundation it is for collaboration.

Together, the community of open source programmers is creating a remarkably active and rich information commons. A study by two scientists at the software company SAP, Amit Deshpande and Dirk Riehle, shows that the commons now contains more than a billion lines of publicly available code, and is growing at a rate of more than 300 million lines per year. Want to add flames to your home movie as a special effect? There are open source software packages for that. Want to control your robotic home telescope? Depending on your telescope, there may well be open source software for that. Open source software is available to do an almost unimaginably broad range of tasks.

The emergence of this rich information commons has radically changed the way programmers work. Before, programmers wrote their programs largely from scratch. Their heroes were people who could, in a few days, whip up a program that would take lesser programmers months to write. To give you the flavor of what skills were valued in those days, consider this story from one of the great pioneers of modern computing, Alan Kay, a recipient of the Turing Award, the highest honor in computer science. It's an admiring story about the programming prowess of Donald Knuth, another legend of computing and Turing Award recipient:

> When I was at Stanford with the [artificial intelligence] project [in the late 1960s] one of the things we used to do every Thanksgiving is have a programming contest with people on research projects in the Bay area. The prize I think was a turkey.
>
> [Artificial intelligence pioneer and Stanford Professor John] McCarthy used to make up the problems. The one year that Knuth entered this, he won both the fastest time getting the program running, and he also won the fastest execution time of the algorithm. He did it on the worst system [...] And he basically beat the shit out of everyone.

Today, programming has changed. Today, a great programmer isn't just someone who can quickly solve a problem from scratch. A great programmer is someone who is also a master of the information commons, someone who, when asked to solve a problem, knows how to quickly assemble and adapt code drawn from the commons, and how to balance that with the need to write additional code from scratch. Such a master can build on the work of others to solve problems faster and more reliably than other lesser programmers. It's a kind of passive collaboration, whose effectiveness grows as the information commons grows. Before they've written even a single line of code, today's programmers are often building on the work of thousands of other programmers. As some programmers like to say: "Good programmers code; great programmers reuse other people's code."

In programming, the information commons took off in the early 1990s, with broad adoption of the internet. But in a more primitive form the ideas of reuse and the information commons were pioneered centuries earlier, in science. When someone publishes a scientific discovery—say, Einstein's famous paper containing the formula $E = mc^2$—other scientists can reuse that result in their own papers, simply citing the original derivation. This allows scientists to build on the earlier work without having to repeat that work. The citation both credits the original discoverer, and provides a link in a chain of evidence. If someone wants to know why $E = mc^2$, they merely need follow the citation to Einstein's original paper. The result is that, as in modern programming, a great scientist isn't merely a person capable of enormously penetrating insights into nature, but one who also has a mastery of the information commons— already published scientific knowledge—and an ability to build on that knowledge. Science is, in this sense, one big collaboration, built on the information commons.

Science's citation-based information commons is powerful, but cumbersome and slow when contrasted with, say, the rapid-fire pattern of reuse in a project such as Wikipedia or Linux. A scientist who used the Wikipedia and Linux pattern—reusing someone else's text word for word, but making a few improvements here and there— would likely receive an indignant note (or worse) from the original author. Yet such improvements are the lifeblood of many online collaborations, enabling extremely rapid iterative improvement, with people focused solely on moving forward, not on rehashing what is already known. A moderately active Wikipedia article may be modified 20 or 30 times by a dozen different people in a week. To get the same cumulative buildup of ideas in many areas of science might take years. Projects such as the Polymath Project speed up the cumulative building process of conventional science, creating a shared space where scientists can rapidly build upon one another's ideas. Citation is perhaps the most powerful technique for building an information commons that could be created with seventeenth-century technology. But as the Polymath Project shows, and as we'll explore in more detail later, modern technologies now enable a better way.

The MathWorks Competition

In 1998, a software company called MathWorks began running a twice-annual computer programming contest that is open to anyone in the world. For each contest MathWorks poses an open-ended programming problem. To give you the flavor of the contest, consider the problem used in the first contest, in 1998, a problem called the CD packing problem: to write a program which, when given a long list of songs, picks out a sublist that comes as close as possible to filling the 74-minute length of a CD. For example, your program might be asked to pick out songs from Pink Floyd's back catalog. You run your program, and it finds a list of songs from the catalog that leaves just 35 seconds of extra space on the CD. But if your program had a better way of selecting songs, you might find yourself with only 15 seconds left on the CD.

The CD packing problem seems artificial. Not too many people have a need to burn CDs that are as close to filled as possible. Despite this, the problem is exactly the kind of challenge many programmers enjoy. It's a simple problem that's easily understood, but can be attacked in many different ways. Like all the MathWorks competitions, the original competition was very popular, attracting more than 100 contestants from all over the world.

Every program entered in a MathWorks contest is given a score reflecting both how quickly the program runs (faster programs get better scores) and how well it solves the problem. In the case of CD packing, programs that came closer to filling the CD were given better scores. Contestants can submit programs at any time during the week-long competition, and are welcome to submit multiple entries, or multiple versions of the same entry. Entries are automatically scored as soon as they're submitted, and the scores immediately placed on a leaderboard. (We'll come back to how the automated scoring is done shortly.) Acclaim goes to people near the top of the leaderboard, however briefly, and so instead of waiting until the end of the week to submit their entries, people submit entries throughout the entire week. The contest's overall winner is the person at the top of the leaderboard at the end of the competition.

What makes the MathWorks competition special is that every time someone submits an entry, the code for their program is immediately made available for other people to download and reuse. That is, anyone can come in, "steal" someone else's code, change it to get an improvement, and then resubmit it as their own, possibly vaulting over the other person on the leaderboard. This ability to reuse other people's code has spectacular consequences. The leading programs are constantly being tweaked by very minor changes, often changing just a single line of code in an earlier entry. Changes come fast and furious, and some contestants become addicted, driven by the instant feedback and the feeling they are just a single idea away from the top of the leaderboard. One contestant has written:

I started to become "obsessed." At home, although I am a father of three children, my full-time job was working on the contest. I worked maybe 10 hours after work each day. On Thursday it was clear that I wasn't going to be able to work seriously (for my job), so I took a day off on Friday.

It's similar to the rapid cycle of feedback that makes computer games addictive. You can always have one more shot at making a tiny improvement. It's arguable whether that's always a good thing—the contestant quoted sounds like he needs to take a holiday from his computer—but this relentless focus also produces amazing results.

The progress of the contest is vividly illustrated by the graph in figure 4.2. The horizontal axis is time, while the vertical axis is the score: for the CD packing problem, lower scores are better. Each dot on the graph represents a competition entry. The scores

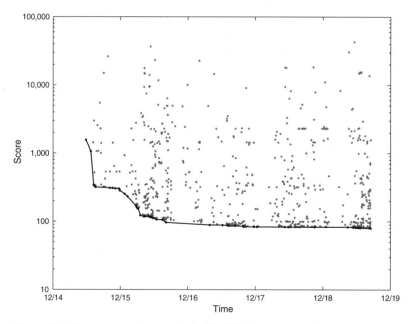

Figure 4.2. The progress of scores in the MathWorks programming competition. Lower scores are better. Credit: Copyright 2011 The MathWorks, Inc. Used by permission. Thanks to Ned Gulley for providing the figure.

dropped so dramatically during the contest that the vertical axis has been rescaled—scores at the top are hundreds of times higher than scores at the bottom. The solid line marks the best score at any given time. As you can see, there are occasional big steps in the line, indicating breakthrough ideas that substantially improve the best score. After such a breakthrough, there is usually a period in which people make many minor tweaks to the leading entry, finding small improvements that further optimize the program, and leave them with the best score.

The difference between the best early entries and the final winner is dramatic. In the CD packing contest, the best early submissions ran quickly, but left six minutes of space on the CD unused. The winning program ran approximately as fast, but left just 20 seconds unused, a nearly 20-fold improvement. It made use of contributions from at least nine people, over dozens of separate submissions. Although it's a competition, the MathWorks contest thus functions in many ways as a large-scale collaboration. The organizer of the contest, Ned Gulley, said of the winning program: "no single person on the planet could have written such an optimized algorithm. Yet it appeared at the end of the contest, sculpted out of thin air by people from around the world, most of whom had never met before." This was not a fluke. The CD packing contest was the first of more than twenty MathWorks competitions that have been held to date. Each contest sees the same gradual emergence of a program whose construction is arguably beyond the ability of any of the individual competitors.

Microcontribution

The Mathworks competition vividly illustrates a pattern that can be used to scale online collaboration: microcontribution. The most common type of entry in the MathWorks competition is an entry that changes just a *single* line of code in some previous entry. That's right, someone comes in and changes just *one* line of code in an earlier entry—very possibly someone else's entry!—and resubmits it as their own. The next most common type of entry changes just two lines. And so on. The result is that even though people are competing, the evolution of the leading entries looks almost like a conversation, with lots of back and forth, as the baton of leadership passes from one participant to another. It's a creative exchange of ideas that drives gradual improvement over time, with different people contributing as best they can.

The same pattern of microcontribution is used in many online collaborations. In Wikipedia the most common edit to an article changes just a single line of that article. In Linux the most common contributions change just a single line of code. A study by two scientists at the software company SAP, Oliver Arafat and Dirk Riehle, showed that this pattern is quite general: in most open source software projects the most common change is to just a single line of code, the second most common change is to two lines, and so on. In the Polymath Project, project leader Tim Gowers asked participants to share just a single idea in each contribution, and to resist the temptation to go off and develop ideas extensively on their own.

Microcontribution lowers the barrier to contribution, encouraging more people to become involved, and also increasing the range of ideas contributed by any particular person. As a consequence it increases the range of expertise available to the collaboration. Recall Yasha, the member of the World Team who contributed the crucial move number 26. Yasha would have been lost playing Kasparov on his own. But it was very helpful, perhaps vital, for the World Team to have access to Yasha's small contribution. Small contributions spark ideas and insight, as people share ideas that they couldn't develop alone, but that can inspire others. If a participant in the Polymath Project or the MathWorks competition was stuck for ideas, they only needed to wait a few hours, watching for new ideas to stimulate and challenge them. Or they could dig into the archives, looking for fresh stimulation from old ideas. Microcontribution thus helps build a vibrant community, a sense that something is afoot, that progress is being made, that even when you, personally, are stymied, other people are moving things forward. Microcontribution is a powerful pattern of collaboration, in short, because the small contributions help the collaboration rapidly explore a much broader range of ideas than would otherwise be the case.

Scores as Signals to Coordinate Expert Attention

I said earlier that entries in the MathWorks competition are scored automatically as soon as they're submitted, but I glossed over how that's done. Imagine you're one of the competition organizers, and one of the competitors has just submitted their program. How should you score it? The obvious thing to do (and the way it's actually done) is to run the program on a few test inputs. You might try it out on (say) three test inputs: the Beatles catalog, a collection of jazz pieces, and a collection of dance music. So on the first run the program would attempt to fill a CD with songs chosen from the Beatles catalog, on the second run it would use songs from the jazz collection, and on the third run songs from the dance collection. You'd then give the program a score determined both by how quickly the program runs and how well it fills up the entire CD on each of the three test inputs. Of course, there's no need for this to be done manually by an organizer. It can all be done automatically as soon as entries are submitted, so the score can be computed immediately. The only caveat is that for this to work the organizers need to keep the test inputs secret—if competitors knew, for example, that their program would be used on the Beatles catalog, they could tailor it specifically to the Beatles catalog, defeating the point of the competition. But provided the organizers are careful to keep the test inputs secret, they can automatically score entries as soon as they're submitted.

Automated scoring is important because the scores help participants focus their attention where it will do the most good. If someone changes a program and causes a big jump (or even a small improvement) in the score, other people notice and check to find out what's been changed: maybe that person has a great new idea. The automated scoring thus makes it easy for programmers to keep tabs on each others' best ideas—even if the number of participants is very large—and to spot opportunities to use their own expertise to make further improvements, and so leapfrog over one another. Some of the programmers, for example, are experts on the detailed ins-and-outs of the programming language (called MATLAB) used in the competition. They watch other people's programs carefully, and use their knowledge of MATLAB to make tiny optimizations, often changing just one or two lines of MATLAB code to be more efficient, and so shaving a fraction of a millisecond off the running time. Other competitors specialize in other ways. Some scour the scientific literature looking for inspiration. Others brainstorm completely new approaches. And some work on hybridizing existing approaches. Amidst all these differing approaches, the automated scoring plays a role similar to prices in a market, providing information that can be used to inform decision making by contest participants. While it's impractical to conduct a conversation involving the more than 100 people who entered in the MathWorks competition—no one has time to pay attention to more than 100 separate voices—the score helps people make good decisions about where to focus their attention, and so fuels rapid improvement.

The MathWorks score is not perfect as a way of coordinating attention. Because the same scoring information is provided to everyone, it leads competitors to concentrate their attention in similar ways. For example, if someone jumps to the top of the leaderboard, then many participants will immediately shift their attention to that entry. Of course, some concentration of attention is good, but if everyone follows the same lead, then the group as a whole may neglect promising directions for exploration. You could imagine more complex signaling mechanisms that would spread attention more widely, and lead to a better allocation of expertise. For instance, people with expertise in optimizing MATLAB code might be directed to programs whose gross structure was changing rapidly, but whose fine detail had not yet been optimized. Or perhaps there could be some way of detecting clusters of programs that make use of similar ideas. Contestants who enjoyed hybridizing different approaches could use this information to help them pick out the best programs in each cluster, and attempt to hybridize those.

These limitations aside, the MathWorks score does a great job of helping coordinate attention, and thus of helping the MathWorks collaboration scale. As a way of directing attention it works much more effectively than, for example, any mechanism available in the Polymath Project, which relied on the acumen of individuals to assess which contributions were worth following up on. It could take hours or days for the polymaths to identify the best new ideas. That's fast, especially when compared to the usual pace of scientific research, but slow compared to the immediacy of the MathWorks score. The situation in Kasparov versus the World was similar to the Polymath Project, although tools such as Krush's analysis tree helped coordinate attention. The better the architecture of attention is at directing attention in this way, the more collective intelligence is amplified.

Converting Individual Insight into Collective Insight

In addition to coordinating attention, the MathWorks score also served the important purpose of helping turn the insights of individual participants into collective insights held by the entire group. Every time someone had an idea that improved a program, this was reflected in their score, making the value of their new idea immediately apparent to all participants. For collaboration to succeed, there must be some way of converting individual insight into collective insight. In other words, the collaboration needs to know what the collaboration knows.

Kasparov versus the World shows what happens when a collaboration only imperfectly converts individual insight into collective insight. As we've seen, the World Team relied on Irina Krush and her colleagues to identify and publicize the best ideas of the World Team. Without Krush's skill at evaluating and comparing analyses, the World Team would likely have done far worse at aggregating the best ideas. Of course, even though Krush and her colleagues put in a mighty effort, their manual approach wasn't as fast or objective as the automated scoring in the MathWorks competition. As a result, much of the available expertise on the World Team was squandered. Many experienced chess players participated on the World Team, and while some enjoyed the experience, others felt alienated, believing their insights were lost in the general noise of discussion. Years after the game, one participant wrote in an online forum:

> If anything in my life that I've participated in that I could label as a perfect example of how a community should NOT solve a problem, it was the KvW match. (which I particpated in heavily and am a master (fide [chess rating] 2276)).

Such disaffection occurred because Krush and a few colleagues were manually integrating the best ideas of thousands of people. Their efforts were remarkable, but of course they could only do the job imperfectly. This caused occasional frustration on the World Team, and almost certainly some missed opportunities. This is a general rule: the more effectively a collaboration can convert individual insight into collective insight, the more effective the collaboration will be.

In fact, the World Team's system for converting individual insight into collective insight broke down badly at a crucial point in the game. As I mentioned earlier, until move 51 the game had seesawed back and forth between Kasparov and the World, with neither side gaining a decisive advantage. By move 51, Kasparov was in a slightly stronger position, and the World Team was fighting for a draw. Unfortunately, at move 51 a member of the World

Team by the name of Jose Unodos claimed to be able to break Microsoft's voting system, and to have stuffed the ballot in favor of a move that he personally liked, but that was not considered a strong move by Krush and most of the other top World Team players. Jose Unodos's preferred move won the vote, the first time since move 9 that Krush's recommendation wasn't played by the World Team. The event helped tip the balance of the game in favor of Kasparov, and damaged the World Team's morale. Eleven moves later, Kasparov won, in a sad end to one of the great games in the history of chess. When a group's ways of converting individual insight into collective insight break down, collective intelligence no longer functions. In the next chapter we'll see that in some fields such breakdowns impose fundamental limits on collective intelligence.

CHAPTER 5

The Limits and the Potential of Collective Intelligence

Collective intelligence is not a problem-solving panacea. In this chapter we'll identify a fundamental criterion that divides problems where collective intelligence can be applied from problems where it cannot. We'll then use that criterion to understand why scientific problems are especially well suited for attack by collective intelligence. To understand the criterion, let's first turn to an experiment done in 1985 by the psychologists Garold Stasser and William Titus. What Stasser and Titus showed is that groups discussing a certain type of problem—a political decision—often do surprisingly badly at using all the information they possess. This perhaps doesn't sound so surprising: after all, everyday political discussion isn't always terribly informative. But what Stasser and Titus showed went much further: group discussion sometimes actively makes people's political decisions worse than they would have been if they had made those decisions individually.

Stasser and Titus began by creating written profiles of three fictional candidates for president of the student government at Miami University, where Stasser was a faculty member. The profiles contained information about the candidates' policies on issues of interest to students—dorm visitation hours, local drinking ordinances, and so on. Stasser and Titus deliberately constructed the three profiles so that one of the candidates was clearly more desirable than the other two. They did this by first surveying students to figure out which traits students found desirable, and then constructing the profiles accordingly. We'll give this extra-desirable candidate a name: we'll call them "Best."

In the first version of the experiment, each student received complete profiles of all three candidates, and was asked to decide who their preferred candidate was. Not surprisingly, 67 percent of the students chose Best. Stasser and Titus then divided the students into small groups of four people each, and asked the groups to discuss which candidate should be president. At the end of the discussion the students were again asked for their preferred candidate. Support for Best increased to 85 percent.

So far, no surprises. But Stasser and Titus also did a second version of the experiment. This time they altered the profiles so that each student received only *partial* information about the three candidates: they removed some of the positive information about Best—things students could be expected to like—and they also removed some of the negative information about one of the undesirable candidates. In fact, any *single* partial profile now suggested that one of the undesirable candidates was actually better than Best. Not surprisingly, when asked to choose a candidate on the basis of these partial profiles, 61 percent of the students preferred the undesirable candidate, while only 25 percent preferred Best. After this, Stasser and Titus again divided the students into small groups of four, and asked the groups to discuss which candidate should be president. But here's the clever bit: when Stasser and Titus were constructing the partial profiles, they were careful to remove *different* information from different profiles, so that each *group* of students would still have *all* the information about all three candidates. Thus each group still had all the information they needed to identify Best as the truly best candidate. Note that the students were warned in advance that not everyone in their group necessarily had the same information about all three candidates.

Now, in this second version of the experiment you'd think Best's percentage would increase after the group discussion, as people shared what they knew and realized that Best was truly the better candidate. But that's not what happened. In fact, after the discussion it was the undesirable candidate whose percentage increased, from 61 percent to 75 percent. Best's percentage actually decreased, from 25 percent to 20 percent. The groups weren't so much sharing information as they were reinforcing the students' preconceived ideas. To put it another way, group discussion didn't make the groups' decisions better, it made them worse. It was a case of collective stupidity, not collective intelligence.

Whats was going on? We've seen many examples showing how groups can use their collective intelligence to perform better than any individual in the group. Yet the Stasser-Titus experiment shows that discussion sometimes makes groups do worse than their average member. Furthermore, the Stasser-Titus experiment is part of a much broader set of findings in group psychology that show that groups—even small groups, or groups of experts—often have trouble taking advantage of their collective knowledge.

For example, in a 1989 follow-up to the original Stasser-Titus experiment, the group discussions were recorded so the experimenters could better understand how the groups came to their decisions. What they found was that instead of exploring all the available information, the groups spent most of their time discussing information they had in *common*. So, for example, if several people all knew that Best held an unpopular position on (say) dorm room visitation, there was likely to be a relatively lengthy discussion of that fact, and the information was likely to be mentioned again later in the discussion. But when someone in the group had a *unique* piece of information about a candidate, a piece of information that only they knew, the discussion of that information was usually perfunctory. That mattered, because in the original Stasser-Titus experiment, negative information about Best was often held in common by several members of the group, while positive information was often held by only a single member.

In 1996 another follow-up experiment was done, this time in a teaching hospital, asking groups to make medical diagnoses on the basis of video clips of patient interviews. Again, the information was partial: each person in the group saw only part of the video interview. The groups making the decisions included three people of different statuses a medical resident, an intern, and a student. Alarmingly, but perhaps not surprisingly, the groups paid much more attention to unique information held by the high-status medical resident. Unique information held by the interns and students was much more likely to be ignored.

These and many other studies paint a bleak picture for collective intelligence. They show that groups often don't do a good job of taking advantage of their collective knowledge. Instead, they focus on knowledge they hold in common, they focus on knowledge held by high-status members of the group, and they often ignore the knowledge of low-status members of the group. Because of this, they don't manage to convert individual insight into collective insight shared by the group. And that's bad news if you're trying to use collective intelligence.

The Limits to Collective Intelligence

Why are projects such as the Polymath Project, Kasparov versus the World, and the MathWorks competition so successful, while the groups in the Stasser-Titus and related experiments perform so poorly? To put it more precisely, why were the groups in the successful projects able to convert their best individual insights into collective insight, while the groups in the Stasser-Titus and related experiments failed to make this conversion? Was the difference due merely to differences in the processes used in the respective cases? Or is there some more fundamental difference, a difference that can't be solved by an improved process, perhaps due to the nature of the problems under discussion?

To answer these questions, I want you to consider a little brain-teaser. I'll give a verbal description of the puzzle, but the puzzle is rather visual, and you may find it illuminating to consult the pictorial explanation given in the picture and caption on the next page. You're given an empty eight-by-eight chessboard, and asked to cover it with one-by-two dominoes, so that only two squares remain uncovered: the square in the bottom left, and the square in the top right. Can you do this? If so, how? If not, why not? You're not allowed to stack dominoes, or break dominoes, or leave dominoes hanging off the edge of the board—everything in the puzzle statement is to be interpreted in the usual way. To further simplify things, we'll also require that each domino covers two adjacent squares on the board—no obliquely placed dominoes are allowed.

Most people don't find this an easy puzzle. But it's worth struggling with it for a few minutes before reading on. If you do, and you try laying out imaginary (or real) dominoes on a chessboard, you'll

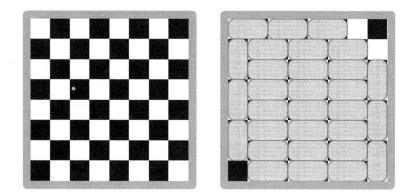

Figure 5.1. The puzzle starts with an empty eight-by-eight chessboard, as shown on the left. You're asked if it's possible to cover the chessboard with one-by-two dominoes, so that only the bottom left and top right squares remain uncovered. On the right, I've shown a failed attempt to do this, which leaves two extra squares in the top right corner uncovered.

discover that no matter how hard you try, you can't quite do it. It's as though there's an unseen obstruction that is somehow preventing you from succeeding. In fact, there is no way of covering the board in the way requested. Here's why. The key is to notice that if you put a domino down on the board, no matter where you put it, it will cover a total of one black square and one white square. So if you put two dominoes down, there will be a total of two black squares covered, and two white squares. Three dominoes means three black squares covered and three white squares covered. And so on. No matter how many dominoes you put down, the total amount of black and white covered will be the same. But notice that both the bottom left and top right squares on the chessboard are black. So to reach a situation where they are the only squares uncovered, you need to somehow cover 32 white squares and 30 black squares. That's an unequal number, so there's no way it is possible.

Although most people find it hard to solve this puzzle, when the solution is explained they quickly say, "Aha, I see it!" It's much easier to recognize the insight that solves the problem than it is to have that insight. Put slightly differently, there's a gap between the difficulty of recognizing the insight and the difficulty of having the insight in the first place. A similar gap is present in examples such as the Polymath Project, Kasparov versus the World, and the

MathWorks competition. Consider the MathWorks competition. It requires tremendous ingenuity to write programs that quickly pack CDs nearly full of songs. But, as we've seen, it's easy to recognize when someone has written a good program: simply run the program on a few test inputs, and check that it runs fast, and leaves little space left over on the CD. It's that gap between the difficulty of writing programs and the ease of evaluating them that fuels collective progress in the MathWorks competition. In chess, recognizing valuable insight isn't quite as straightforward, but a competent chess player such as Krush can recognize and understand an exceptionally insightful analysis of a particular position, even if she couldn't have come up with the analysis on her own. The best analyses may even stimulate the same feeling of "Aha, how clever!" as in the domino puzzle. Krush can't play consistently at Kasparov's level, but she's good enough to recognize when other people are (momentarily, at least) playing at that level, and to understand their analyses. And in the Polymath Project, participants could recognize when others had mathematical insights that exceeded their own, and could incorporate those insights into their collective knowledge. Again, it's that "Aha!" feeling stimulated by a clever insight. Each project thus has used this gap between our ability to have and to recognize useful insights, in order to convert individual insight into collective insight.

The problem in the Stasser-Titus experiments is that the small group discussions did not reliably convert individual insight into collective insight. Intellectually, many of the students participating in the experiments would no doubt have agreed that the way to go was to systematically pool all their information, and then to make a decision based on the combined profiles so constructed. But in practice they didn't do that. And, given the context, this is not surprising. In everyday political discussion, most of us don't assess politicians by building up a complete picture of their positions. We're too busy figuring out how their positions relate to our values and our interests.

Suppose, however, that the groups in the Stasser-Titus experiment actually had begun their discussion by systematically pooling all their information. That experiment has never, to my knowledge, been done, but I think we can be sure it would dramatically change the outcome. So the problem in the Stasser-Titus groups was in part a failure of process; an improved process would result in dramatically better outcomes. But it wasn't solely a failure of process. Even if the groups had systematically shared information, different students would still have had unresolvable differences of opinion. If one student loves to drink and party, while another strongly opposes drinking on religious grounds, then they may never agree on political choices, no matter how good the process that is being used.

This points the way to a fundamental requirement that must be met if we're to amplify collective intelligence: participants must share a body of knowledge and techniques. It's that body of knowledge and techniques that they use to collaborate. When this shared body exists, we'll call it a *shared praxis*, after the word *praxis*, meaning the practical application of knowledge. Whether a shared praxis is available determines whether collective intelligence can be scaled up, or whether it cannot be.

As an example of a shared praxis, imagine a large group is working together on the domino problem. As soon as any single person in the group finds that the domino problem is impossible to solve, they can quickly convince the others, because each step in their reasoning is so self-evidently correct: we all share the same basic reasoning skills. That's an example of a shared praxis. In a similar way, there's a shared praxis for work in mathematics—all the standard methods of mathematical reasoning, and norms about mathematical discourse—and that's why participants in the Polymath Project could recognize and agree on when mathematical progress was being made. Also similarly, the score in the MathWorks competition implicitly defined a shared praxis: any change to a program that improved the score was understood by participants to be progress. In chess, the shared praxis isn't as strong as it is in mathematics and computer programming: even top chess players sometimes disagree about the value of different analyses. Nonetheless, there is a large body of chess knowledge that is broadly agreed upon by strong players, and this shared knowledge means that the stronger players on the World Team could usually agree on which analyses were best.

Those are all examples of problems where there is a shared praxis. But for many problems there is no shared praxis. For instance, as we've seen, there is no strong shared praxis available in politics. People can easily disagree over basic values. And if a group doesn't have such a shared praxis, then disagreements will arise that can't be resolved. Once an unresolvable disagreement arises, the community will begin to fragment around that disagreement, limiting the ability to scale up collaboration. Now, for one-off problems—say, the problem of guessing the weight of an ox at an English country fair (see page 7)—that maybe doesn't much matter. But for working together through multiple stages to solve a problem, such fragmentation imposes fundamental limits on the scale of collaboration.

Politics is just one of many fields that lack a strong shared praxis. The same is also true in many of the fine arts, where assessing creative works is often highly contentious. For instance, to decide which of two paintings is better, we make use of our own aesthetic standards, standards that may be quite different from those held by other people. Similarly, we may reasonably disagree over which of two musical compositions is better. This isn't to say that there is no notion whatsoever of an objective standard in the arts. Pretty much everyone agrees that the Beatles are better than some random boy band. But in comparing the Beatles to Bach, reasonable people may disagree. In making that statement I've no doubt offended music snobs all over the world. But the point is that I've offended both the classical music snobs, who can't believe the Beatles begin to compare to Bach, and also the pop music snobs, who believe that Bach belongs to a tradition that has since been surpassed. When such traditions coexist, it is extremely difficult for people in the two traditions to collaborate, because they have no basis to agree on when they're making shared progress. This isn't a negative judgment about such fields—great musicians, painters, and politicians all operate near the limits of human ability—but it is an important limitation on when collective intelligence may be used.

It's not just politics and the fine arts that don't have a strong shared praxis. Many academic fields lack one as well. Think of criticism of English literature. Critics are not going to one day put down their quills and arrive at a common understanding of Shakespeare. Indeed, arriving at such a common understanding isn't the point. In such fields a plurality of points of view is a feature, not a bug, and a new way of understanding Shakespeare is to be celebrated. But this same plurality of points of view makes it difficult to recognize and integrate the best insights from a large group of people. Any such attempt at collaboration inevitably gets bogged down in discussions about basic values, and questions about what makes a contribution worthwhile. In such fields, agreement doesn't scale, and that severely limits our ability to convert individual insight into collective insight, and so prevents application of collective intelligence.

Some fields sit near the cusp dividing fields where it's feasible to scale collective intelligence and those where it is not. In economics, for example, there are many powerful methods of reasoning that are agreed upon by most economists: an understanding of how trade can make everyone better off, the idea that printing more money usually causes inflation, and so on. But economists *don't* agree on some of the most fundamental questions of economics. As the old joke goes, if you put five economists in a room, they'll give you six wildly differing opinions. US President Harry Truman is supposed to have asked for a one-armed economist, one who couldn't say "On the other hand." So while there is a shared praxis in economics, it's not as strong as the shared praxis in fields such as mathematics, computer programming, and chess. As a result there are many questions in economics that can't be attacked by the methods of collective intelligence. It's only in a few parts of economics, such as the study of some mathematical models of finance and the economy, where a strong shared praxis is available. It's in those parts of economics where collective intelligence can be scaled.

The availability of a shared praxis isn't the only challenge in applying collective intelligence. There are many other practical problems. An example is groupthink, where members of a group may be more interested in getting along with one another than in critically evaluating ideas. Or groups may become echo chambers, with group members merely reinforcing each others' existing opinions. In some groups, basic norms of civil behavior breakdown. This sort of breakdown has destroyed many open source software collaborations, and bedevils many badly designed forums on the web, which may become havens for internet trolls, and other antisocial behavior. The projects we've discussed have overcome these and similar problems: some have succeeded with flying colors (the Polymath Project), while others just barely succeeded (World

Team deliberations sometimes teetered on the edge of breakdown because of lack of civility). Similar problems also afflict offline groups, and much has been written about the problems and how to overcome them—including books such as James Surowiecki's *The Wisdom of Crowds*, Cass Sunstein's *Infotopia*, and many other books about business and organizational behavior. While these practical problems are important, they can often be solved with good process. But no matter how good the process, there remains a fundamental dividing line: whether a shared praxis is available. In fields where a shared praxis is available we can scale collective intelligence, and get major qualitative improvements in problem-solving behavior, such as designed serendipity and conversational critical mass. For fields without a shared praxis, online tools don't give us the same qualitative shift.

The Shared Praxes of Science

Science is well suited for collective intelligence. Most fields of science have a large repository of powerful techniques shared by the scientists working in that field. There are widely agreed standards for what it means for an argument or analysis or experimental procedure to be correct. This was illustrated vividly by the Polymath Project, where discussion was carried out in a remarkably civil tone. On those rare occasions where disagreement occurred, it was usually because someone had made an outright error in reasoning. Someone else would point out the error, without rancor, whereupon the originator would immediately acknowledge their mistake. This is not to say that participants never engaged in speculation, but they carefully marked their speculation as such, and didn't present it as incontrovertible fact. On nearly all crucial issues the participants rapidly agreed on when a line of argument was right and when it was wrong, and on when an idea was promising and when it was not. It was that rapid agreement which made it possible to scale up collaboration.

As an illustration of how strongly held these standards are in science, consider the work of the young Albert Einstein, not the scientific icon we know of today, but as an unknown 26-year-old clerk working in the Swiss patent office, unable to find a job as a professional physicist. From that position of obscurity, in 1905 Einstein published his famous papers on special relativity, radically changing our notions of space, time, energy, and mass. Other scientists had partially anticipated Einstein's conclusions, but none so boldly and forcefully laid out the full consequences of special relativity. Einstein's proposals were astounding, yet his arguments were so compelling that his work was published in one of the leading physics journals of his day, and was rapidly accepted by most leading physicists. How remarkable that an outsider, a virtual unknown, could come in to challenge many of our most fundamental beliefs about how the universe works. And, in no time at all, the community of physicists essentially said, "Yeah, you're right."

As another example, consider the discovery of the structure of DNA. This discovery was made by James Watson and Francis Crick, using data due in part to Rosalind Franklin. All three were young, unheralded scientists: Watson was 24, and Crick was 36, reestablishing himself after a brief career in physics and work in the British Admiralty during World War II. Franklin was 32. Racing them to the discovery was the world's leading chemist, Linus Pauling. More than a decade earlier, the brilliant Pauling had made a series of discoveries that would eventually win him a Nobel Prize in Chemistry. If he could solve the structure of DNA, another prize would surely follow. At one point during the race he gave Watson and Crick a tremendous scare, announcing that he'd found the structure. But Watson and Crick spoke with Pauling's son, Peter Pauling, who showed them Pauling senior's proposed structure for DNA. To their astonishment, they quickly realized that Pauling was wrong: the world's greatest chemist had made a simple mistake in basic chemistry, a mistake his own textbooks should have alerted him to. Watson and Crick went back to their work with renewed intensity, and soon after found the right structure. When that happened it didn't matter that Pauling was world famous while Watson, Crick, and Franklin were unknowns. The scientific community rejected Pauling's work, and hailed the double helix as one of the scientific discoveries of the century.

The examples of Einstein and of Watson, Crick, and Franklin illustrate the strength of the shared praxis in science. To an extent unusual in many parts of life, in science it's often the person with the best evidence and best arguments who wins out, and not the person with the biggest reputation and the most power. Pauling may have been widely acknowledged as the world's leading chemist, but other chemists could see just as surely as Watson and Crick that Pauling's structure was simply wrong. This strong shared praxis makes science well suited to collective intelligence.

This strong shared praxis doesn't mean that science is a clean and simple process. The actual day-to-day process of doing science is messy and speculative and filled with error and argument. The scientist Richard Feynman was so full of irrepressible brainwaves and "great" ideas, most of which later proved to be wrong, that according to his biographer James Gleick his cannier colleagues developed a rule of thumb: "If Feynman says it three times, it's right." The same could be said for many scientists. Often a scientist begins an investigation with little more than a whiff of an idea, a suspicion that some hypothesis is true. They sketch out a way of testing it, often vaguely at first, gradually filling in more and more details. Experiments often need to be performed many times, with the experimental design gradually changed and improved, as the scientist understands better what evidence is required in order to be convincing. All this is a slow process that involves lots of speculation and argument and false starts, as the scientist gradually moves to more and more robust arguments and evidence. The end goal, though, is a set of arguments and evidence that adheres to the shared praxis of the field. And that is quite unlike a discussion of Bach-versus-the-Beatles, or a political discussion, or a discussion of Shakespeare, where in the end there may remain a fundamental division over basic values. Of course, scientists do still sometimes publish wrong or mistaken or unconvincing papers. But even when a scientist publishes such a result, other scientists can go back and repeat the experiments to find flaws, or point out shortcomings in the arguments. In short, they can retest the results against the shared praxis of the field, and find them wanting. It's this ability to be wrong in a clear-cut way that enables forward progress. In this sense science is, as I said earlier, already one big collaboration, held together by common standards of evidence and reasoning.

Are there parts of science without a shared praxis, parts more like economics, say, where the problems are so challenging that the field is still a proto-science, with shared knowledge and techniques only starting to emerge? As an example, one of the big open problems of physics is the problem of finding a quantum theory of gravity—a single theory that unifies both quantum mechanics and Einstein's theory of gravity. It's one of the toughest problems of physics, a problem that has defeated the best minds for decades. In the 1980s an approach to the problem known as string theory rose to prominence, and gradually came to dominate work on quantum gravity. At the same time, a much smaller number of physicists continued to pursue other approaches to quantum gravity. In recent years a debate called by some the "string wars" has been waged between advocates of the different approaches. Many physicists claim string theory is the only reasonable approach to quantum gravity. Others, including Stephen Hawking, Roger Penrose, and Lee Smolin, believe different approaches are worth pursuing. Remarkably, some prominent string theorists dismiss the non-string theorists not just as wrong, but as misguided, or even as fools. When such a fundamental division occurs, it is nearly impossible for large groups to collaborate across that division. Collective intelligence can only be applied within the respective tribes, where there is a shared praxis. And such collaborations need to be guarded carefully against disruption by the rival tribe.

The situation in quantum gravity is unusual. In most areas of science, scientists can compare two competing explanations of a phenomenon to an experiment, and realize that one explanation is right (or, at least, not ruled out by the experiment), and the other is wrong. Or a scientist can point out a hole in another's experimental procedure, and everyone will agree that, yes, that really is a hole, it doesn't come up to the expected standard. But in quantum gravity the phenomena being studied are so remote that we don't yet know how to do experiments—it's still all theory. And developing the basic theory is so challenging that picking out starting assumptions has become to some extent a matter of personal taste, in a manner similar to the fine arts. It's these highly unusual conditions that have prevented the development of a shared praxis. By contrast, in most other fields of science, there is a strongly held shared praxis. And so science gives us a marvelous opportunity to amplify our collective intelligence.

Using Collective Intelligence in Science

In part 2 of this book, we've seen how online tools can be used to amplify collective intelligence, both making groups smarter and making smarter groups. As we come to the end of part 1, let's use those ideas to imagine some of the ways online tools could be used to amplify collective intelligence in science. We'll take a personal point of view, trying to imagine a few of the ways these tools might impact the day-to-day life of an individual scientist. In the chapters to come we'll see how some of these dreams are being realized and even exceeded today. We'll also see how other parts of these dreams are blocked by current social practices within science—and how that can be changed.

Imagine it's a few years in the future, and you're a theoretical physicist working at the California Institute of Technology (Caltech), in Pasadena. Each morning you begin your work by sitting down at your computer, which presents to you a list of ten requests for your assistance, a list that's been distilled especially for you from millions of such requests filed overnight by scientists around the world. Out of all those requests, these are the problems where you are likely to have maximal comparative advantage. Today, one of the requests immediately catches your eye. A materials scientist in Budapest, Hungary, has been working on a project to develop a new type of crystal. During the project an unanticipated difficulty has come up involving a very specialized type of problem: figuring out the behavior of particles as they hop around randomly ("diffuse") on a triangular latticework. Unfortunately for the materials scientist, diffusion is a subject they don't know much about. You, in turn, don't know much about crystals, but you are an expert on the mathematics of diffusion, and, in fact, you've previously solved several research problems similar to the problem puzzling the materials scientist. After mulling over the diffusion problem for a few minutes, you're sure that the problem will fall easily to mathematical techniques you know well, but which the materials scientist probably doesn't know at all.

You message the materials scientist with an outline of a solution to their problem. Over the next few days you communicate back and forth, jointly fleshing out a solution, filling in many details, and translating your mathematical ideas into the language of materials science. Much work on the original project remains to be done, but a critical bottleneck has been overcome. Your reward is a happy collaborator, eventual coauthorship on a paper, and the pleasure of learning a little about the physics of crystals and how it relates to your expertise in diffusion. Your collaborator's reward is to save hundreds of hours they otherwise would have spent becoming expert enough to solve the diffusion problem. The community as a whole is also rewarded: with your help the problem was solved much faster and at lower cost than would otherwise have been the case, the scientific results obtained are stronger, and the explanation of the results in the published paper is clearer. Everyone benefits because of your comparative advantage—you have the skills to make short work of a problem that would take the materials scientist weeks to solve. You each get to do what you're best at—and society saves thousands of dollars.

On the same morning that all this begins, you notice another striking request on your list of top-ranked requests. It comes from a student in Bangalore, India, who wants some help learning about recent research on using computer algorithms to simulate complex quantum systems. They don't know any local experts, and are learning from online papers, which they find confusing at some points. You've received the request because you're an expert on such algorithms, and can easily answer the student's questions. Furthermore, you've asked your system to alert you to a few student requests for assistance each week, tailored to areas where you have a special expertise. A rapid-fire exchange with the student ensues over the next couple of days, clearing up much of their confusion. Your work with the student is automatically noted in an archive of your scientific activity, along with statistics showing your contribution to public outreach.

A few other requests also show up in your list of top-ranked requests, but you decide you don't have time to help out. Among these are several more collaboration requests broadly similar to that from the materials scientist, although differing in the details;

a request for assistance from a local school; and a request for reading material from a student whose thesis topic overlaps with several of your old papers. All of these requests will be seen by tens or hundreds of other people, most of whom, like you, have a special expertise closely related to the requests. Response is voluntary, and none of the requests are directed only to you.

All this is made possible by a ranking algorithm that prioritizes the millions of requests for assistance made daily so that you see only the requests where you personally are likely to have the greatest interest and the greatest comparative advantage. The ranking algorithm takes into account your areas of expertise, what requests you've responded to in the past, the history of the people making the requests, and preferences such as your desire to help students. By judiciously selecting requests, you can maximize the impact of your work.

Around the world, similar patterns are being repeated millions of times over. A cognitive scientist in Ottawa is trying to replicate an experiment showing how a particular optical illusion can be suppressed by changing the color of some parts of the illusion. When she began work, she tried to figure out how to replicate the experiment just from a broad understanding of the original experiment. She made good progress, but occasionally got stuck, whereupon she consulted online videos showing the experiment being done in two other laboratories. That helped, but she's still having trouble reproducing the results. After several days of being bogged down, last night she sent out a request for help, hoping to find someone with expertise both in optical illusions and in how the nervous system combines the color information coming from the different cones in the eye. This morning, she's heard from a psychophysicist in Iowa, who's sent along a modified color scheme, and some instructions on how to recalibrate the color scheme, if necessary. In short order she solves the problem, and the experiment is up and running.

Meanwhile, in a research lab in Shanghai, China, a biologist is working late at night, genetically sequencing a strain of the influenza virus. When he's done with the sequencing, he queries online databases to compare the virus's genetic makeup to all known viral strains. He discovers that this is, as he suspected, a new variation of influenza. Over the next few weeks he will design a vaccine for the new virus. To design the vaccine, he uses software that pulls down information from dozens of online databases, effectively asking and receiving answers to thousands of questions about viruses, their genes, the proteins they produce, and the effect of those proteins. But unlike our earlier examples, these questions aren't made as one-off requests for information. Instead, the software is asking the questions and receiving the answers in an automated way, almost invisibly to the scientist, weaving together knowledge acquired by tens of thousands of biologists, and then recombining that knowledge to help make a new discovery.

All over the world millions of connections like these are being made. Scientists whose work is currently stymied by difficult scientific problems are being connected to other scientists who have the expertise to quickly solve those problems. It's an online market in expert attention, a sort of collaboration market that makes everyone more efficient and capable, better able to work on problems where they have a comparative advantage, and leaving other work for other people. In this collaboration market the sort of connections that today only happen by serendipity instead happen by design. At the same time as these connections between scientists are being made, a quieter but far greater exchange of knowledge is going on in the background, as scientists download and process vast quantities of data, in this way taking advantage of knowledge previously acquired by thousands of other scientists. This, too, is a collaboration market, but instead of specialized, one-off questions, it is for questions so standardized that they can be answered automatically.

Let us zoom back to the personal level, back to Pasadena and Caltech. Aside from your new collaboration with the materials scientist in Hungary, you spend most of your day working on one of your ongoing projects, an ambitious undertaking to design a quantum computer. Quantum computers are hypothetical computers that harness quantum mechanics to solve problems that aren't feasible to solve on conventional computers. While large-scale quantum computers promise to be remarkable devices, building them is a huge challenge, because quantum states are very delicate. To meet this challenge, six months ago you and two colleagues started a project to design a quantum computer that really can be scaled up.

Your project involves a special approach to quantum computing called topological quantum computing, an approach that relies on insights from many different fields of science, ranging from the mathematical field of topology to the physics of superconductors, and from semiconductor fabrication to all the detailed ins and outs of the theory of quantum computing. The project has rapidly grown to involve more than 100 scientists, from all over the world, collaborating online. Some of those scientists are theorists, with diverse expertise ranging across the many areas involved. But most are experimentalists, including some of the world's top experts on superconductors and semiconductors, as well as materials scientists who specialize in preparing high-quality material samples. Those experimentalists are sharing their tricks and tips about what's possible in the most advanced laboratories, the type of folklore knowledge that separates the labs at the forefront from those a step behind.

The collaboration hasn't always made smooth progress toward its goal. But even when apparently insurmountable obstacles have arisen, it's often been possible to get past those obstacles using the same collaboration market that saw you begin your Hungarian collaboration this morning. This has also helped draw new people and new expertise into the collaboration. As the collaboration has grown, it's become your biggest ongoing commitment, and most days you spend at least an hour or two on the project. It's gone much further than you first imagined, as the collaboration has found its way around obstacles that you thought were impassable, and, as the ideas of the collaboration move from the speculative to the more feasible, some of the labs involved are beginning to prototype some of those ideas.

Some readers—especially, perhaps, those who have worked as scientists—may read the above paragraphs and think they sound like a pipe dream. "Why," they may ask, "would those experimentalists ever help one another in this way? In the real world, they'll never share the key ideas that are their competitive advantage." Today, this is true, and we'll return to this problem in different guises repeatedly in the coming chapters. But as our understanding deepens, we'll see that while it is a challenging problem, it's not insurmountable. For now, though, we'll defer discussion.

These are just a few ideas to stimulate your thinking about how online tools and collective intelligence can be used to change science. Of course, far more is possible. Imagine completely open source approaches to doing research. Imagine a connected online web of scientific knowledge that integrates and connects data, computer code, chains of scientific reasoning, descriptions of open problems, and beyond. That web of scientific knowledge could incorporate video, virtual worlds, and augmented reality, as well as more conventional media, such as papers. And it would be tightly integrated with a scientific social web that directs scientists' attention where it is most valuable, releasing enormous collaborative potential.

In part 2 of this book we'll explore, in concrete terms, how the era of networked science is coming about today. We'll see, for example, how vast databases containing much of the world's knowledge are being mined for discoveries that would elude any unaided human. We'll see how online tools enable us to build new institutions that act as bridges between science and the rest of society in new ways, and that can help redefine the relationship between science and society. The place where these ideas are being most fully realized is in basic science, and so the focus in part 2 is on basic science—by contrast, applied science is often carried out by small groups working in secret, inside private companies, and that secrecy limits their ability to scale up collaboration. But even in basic science, there are serious obstacles to be overcome. Simple ideas such as collaboration markets, open source wiki-like research papers, and sharing of data and computer code face considerable cultural obstacles. We'll develop the idea that for networked science to reach its full potential, it must be *open science*, based on a culture in which scientists openly and enthusiastically share all their data and their scientific knowledge. And, finally, we'll see how that more open scientific culture can be created.

PART 2

Networked Science

CHAPTER 6

All the World's Knowledge

Don Swanson seems an unlikely person to make medical discoveries. A retired but still active information scientist at the University of Chicago, Swanson has no medical training, does no medical experiments, and has never had a laboratory. Despite this, he's made several significant medical discoveries. One of the earliest was in 1988, when he investigated migraine headaches, and discovered evidence suggesting that migraines are caused by magnesium deficiency. At the time the idea was a surprise to other scientists studying migraines, but Swanson's idea was subsequently tested and confirmed in multiple therapeutic trials by traditional medical groups.

How is it that someone without any medical training could make such a discovery? Although Swanson had none of the conventional credentials of medical research, what he did have was a clever idea. Swanson believed that scientific knowledge had grown so vast that important connections between subjects were going unnoticed, not because they were especially subtle or hard to grasp, but because no one had a broad enough understanding of science to notice those connections: in a big enough haystack, even a 50-foot needle may be hard to find. Swanson hoped to uncover such hidden connections using a medical search engine called Medline, which makes it possible to search millions of scientific papers in medicine—you can think of Medline as a high-level map of human medical knowledge. He began his work by using Medline to search the scientific literature for connections between migraines and other conditions. Here are two examples of connections he found: (1) migraines are associated with epilepsy; and (2) migraines are associated with blood clots forming more easily than usual. Of course, migraines have been the subject of much research, and so those are just two of a much longer list of connections that he found. But Swanson didn't stop with that list. Instead, he took each of the associated conditions and then used Medline to find further connections to that condition. He learned that, for example, (1) magnesium deficiency increases susceptibility to epilepsy; and (2) magnesium deficiency makes blood clot more easily. Now, when he began his work Swanson had no idea he'd end up connecting migraines to magnesium deficiency. But once he'd found a few papers suggesting such two-stage connections between magnesium deficiency and migraines, he narrowed his search to concentrate on magnesium deficiency, eventually finding eleven such two-stage connections to migraines. Although this wasn't the traditional sort of evidence favored by medical scientists, it nonetheless made a compelling case that migraines are connected to magnesium deficiency. Before Swanson's work a few papers had tentatively (and mostly in passing) suggested that magnesium deficiency might be connected to migraines. But the earlier work wasn't compelling, and was ignored by most scientists. By contrast, Swanson's evidence was highly suggestive, and it was soon followed by therapeutic trials that confirmed the migraine-magnesium connection.

If you suffer from migraines you'll know the discovery of the migraine-magnesium connection hasn't resulted in a cure or a surefire treatment. Today, magnesium deficiency is just one of many factors known to contribute to migraines, and the primary cause of migraines remains elusive and the subject of debate. Nevertheless, uncovering the migraine-magnesium connection was a significant step in understanding what makes migraines happen and how to stop them. Furthermore, the significance of Swanson's work goes well beyond medicine. While it has become the conventional wisdom of our age to bewail the information explosion, as though the massive increase in our knowledge is somehow a bad thing, Swanson tipped this point of view on its head. He saw the growth of knowledge not as a problem, but as an opportunity. He realized that tools such as Medline expand our ability to find meaning in humanity's collective knowledge, and so enable us to discover patterns in the whole that are invisible to unaided humans. No human mind could ever encompass the millions of experiments indexed by Medline. Fortunately, no one mind needs to. Working in symbiosis with tools such as Medline, we can extend our minds so that we can find connections hidden in superhuman amounts of knowledge. Effectively, such tools are enabling a new method of scientific discovery.

Searching for Influenza

The method used by Swanson to discover the migraine-magnesium connection is just one of many new ways of finding meaning hidden in existing knowledge. A different approach has recently been used by scientists at Google and the US Centers for Disease Control and Prevention (CDC) to develop a better way of tracking the spread of the influenza virus—the flu. Each year, the flu kills between 250,000 and 500,000 people around the world. Governments and health organizations carefully track the spread of the flu, so they can respond quickly to outbreaks, and prevent pandemics such as the 1918 Spanish flu, which killed more than 50 million people. In the United States, the flu is tracked by the CDC, which signs up doctors across the country to participate in a tracking program. When a patient reports flu-like symptoms—a fever and sore throat or cough—the doctor reports that visit to the CDC. Only a small fraction of doctors participate in the CDC program, but enough do to allow the CDC to build up an accurate regional and nationwide picture of the flu. When an outbreak occurs, the CDC can mobilize, stepping up vaccination programs in the region and getting the word out in the media. But a problem with the system is that it takes one to two weeks for cases of the flu to show up in CDC reports. That time lag is a serious concern, because flu outbreaks can grow rapidly in just a few days.

Hoping to speed up the CDC's system, the Google and CDC scientists wondered if search queries entered by users into Google's search engine could be used to instantaneously track where the flu is occurring. The idea is that if there's a surge of people in the city of Atlanta searching for (say) "cough medicine," chances are there's been an increase of flu in Atlanta. To get good results, the Google and CDC scientists took the CDC's historical flu data from 2003 to early 2007, and looked for correlations with common Google search queries. They found 45 search queries that were especially well correlated with the historical flu data. Using those queries they built a model that they hoped could be used to instantly figure out where the flu is occurring, just by monitoring Google searches. They then tested that model by comparing it with a new set of data, the CDC data from the 2007–08 flu season. Their model gave nearly perfect (97%) agreement! In other words, Google's search queries can be used to determine where flu outbreaks are happening, and how large they are, but without the time lag suffered by the CDC. What's more, Google search queries can be used to track influenza not only in the United States, but anywhere large numbers of people are using Google, including places where there is no CDC-like organization tracking disease. Google has built a website called Google Flu Trends that uses search queries to track influenza in 29 countries.

The Google Flu Trends results require a couple of caveats. First, many doctors in the United States now use electronic medical record-keeping systems, and the CDC has recently partnered with the makers of one of those systems, General Electric, to develop a new tracking system that should give it a near real-time ability to track reports of influenza from 14 million patients. It's possible and perhaps likely that the CDC's new system will obsolete Google Flu Trends, at least in the United States. Second, the CDC data used to build the Google-CDC system did not, strictly speaking, track influenza. Rather, it tracked "influenza-like" illnesses from reports of symptoms such as cough and sore throat that are often associated with the flu. Other conditions such as colds can produce similar symptoms. A follow-up study done in 2010 confirmed that, not surprisingly, Google Flu Trends is significantly better at tracking influenza-like illnesses than it is at tracking actual laboratory-confirmed cases of influenza. It's a helpful diagnostic tool, not a perfect way of tracking the flu.

Using Google to predict the flu is interesting, but even more interesting are the other possibilities it suggests, possibilities that go beyond medicine and into every aspect of life. Follow-up research has already shown that search queries can be used to predict trends in unemployment and in housing prices, and even to predict how well songs will do on the music charts. What else might be possible? Could Google figure out which search queries predict changes in the stock price of some company, say, Microsoft? What about the behavior of the Dow Jones Industrial Average? Or which technology startup is the best target for acquisition? Or the outcome of the next US presidential election? Or a coup d'état in an unstable country? Suppose Google was tracking the searches of law clerks working at the US Supreme Court—might it be possible to predict court decisions? Or perhaps to figure out what concerns an individual justice has while a case is being heard? Suppose a Google user is making searches that suggest they're planning a bank robbery. Should Google notify law enforcement officials? At a media conference in Abu Dhabi in 2010, Google CEO Eric Schmidt said, "One day we had a conversation where we figured we could just try to predict the stock market. And then we decided it was illegal. So we stopped doing that." It's difficult to know whether to be reassured or horrified. Of course, it's not just Google that's in a position to do this kind of data mining. Many other organizations—banks, credit card companies, and popular websites such as Facebook and Twitter—have access to data sources that may be used to understand and even predict human behavior. If you have access to data and the means to make sense of it, data is power.

Finding Meaning in All the World's Knowledge

For nearly all of recorded history, we human beings have lived our lives isolated inside tiny cocoons of information. The most brilliant and knowledgeable of our ancestors often had direct access to only a tiny fraction of human knowledge. Then, in the 1990s and 2000s, over a period of just two decades, our direct access to knowledge expanded perhaps a thousandfold. At the same time, a second, even more important expansion has been going on: an expansion in our ability to find meaning in our collective knowledge. We see this expansion in Swanson's use of Medline to find connections hidden in our collective medical knowledge, or the way Google and the CDC combined the CDC's existing (but inadequate) knowledge of reported flu with Google's search data, to figure out a better way of tracking the flu. We also see examples in our everyday lives, such as Google's ability to answer our questions, finding just the right webpage, news article, scientific paper, or book. Tools such as Google and Medline redefine our relationship to knowledge, by giving us ways of finding previously hidden meaning, all the "unknown knowns" that are implicit in existing human knowledge, but that are not yet apprehended because of the massive scale of that knowledge. Earlier in this book we saw how collective intelligence can be amplified by restructuring expert attention, to take better advantage of the available expertise. In this chapter we'll discuss a complementary approach to amplifying collective intelligence: to build tools that perform cognitive tasks directly, operating on knowledge itself, by searching for meaning and hidden connections in our collective knowledge.

The remainder of this chapter is in two parts. The first part tells the story of a project from astronomy called the Sloan Digital Sky Survey (SDSS). The SDSS is surveying the universe, much as early mapmakers surveyed the Earth, using a robotic telescope to explore the sky broadly, so far taking images of 930,000 galaxies. Those images aren't just pretty pictures; they're being mined by astronomers to answer all sorts of questions about our universe. We'll learn how the SDSS has been used to find the biggest known structure in the universe, a giant chain of galaxies 1.37 billion light-years long; to discover new dwarf galaxies near our Milky Way galaxy; and to find a pair of orbiting black holes. But although these discoveries are fascinating in their own right, there's a deeper reason we're interested in the SDSS. That's because although access to human knowledge has expanded enormously over the past two decades, a great deal of scientific knowledge *isn't* yet publicly accessible, and a struggle is going on to make it more accessible. And so the first part of the chapter tells the story of the expansion of the information commons in science, using the SDSS as a concrete example to understand both the benefits and the challenges of that expansion. That concrete understanding prepares us for the second part of the chapter, where we broaden our focus to think about the big picture. What are the implications of making all the world's knowledge openly available? And what new methods of discovery will it enable?

Exploring the Digital Universe

The largest known structure in the universe is a chain of galaxies called the Sloan Great Wall. It's 1.37 billion light-years long, contains thousands of galaxies, and is about 1 billion light-years away from Earth. That's so far away that those galaxies are too faint to see with the naked eye, but if you could see them, the Sloan Great Wall would stretch across nearly a third of the sky, all the way from the constellation of Virgo, through Leo, and on to Cancer. It's a pretty sight to imagine, all those galaxies twinkling across the night sky!

The Sloan Great Wall was discovered in 2003, when a team of eight scientists, led by J. Richard Gott III of Princeton university, decided to make a visual map of the entire known universe. This sounds grandiose, but they did it for the same reason we make maps of cities and countries: displaying our knowledge visually can make it easier to understand what we know. Imagine how difficult geography would be if we didn't have maps, but instead had to rely entirely on verbal descriptions. Problems that are easy to solve visually, like figuring out how many continents there are, would all of a sudden become difficult research problems. One imagines early geographers holding research conferences on "Resolving the Number of Continental Land Masses," perhaps with fierce arguments about questions such as whether Asia and North America are truly separate continents.

A big difficulty in making a map of the universe is knowing what's out there. Modern telescopes let us see trillions of objects, but for the most part astronomers concentrate on looking at just a tiny fraction of those objects. This perhaps sounds surprising, but imagine you're an astronomer: wouldn't you prefer to spend your time observing something you already know is extremely interesting, such as the supermassive black hole in the core of the Milky Way galaxy, instead of some random star in some random galaxy? Most astronomers thus spend most of their time looking at objects already known to be interesting. It's like the difference between exploring a city broadly to find interesting new places, versus the temptation to only revisit familiar haunts. To find interesting new objects in the sky, someone needs to strike out and explore the sky broadly.

This is where sky surveys come in. Instead of looking in exhaustive detail at known objects, the telescopes used in sky surveys systematically scan the whole sky, building up a broad picture of the universe. Sky surveys are the foundation of astronomy, often giving us the first clues about which objects to look at in more depth. One of the earliest sky surveys was the *Almagest*, written by the astronomer Ptolemy of Alexandria in the second century CE. Ptolemy didn't have a telescope, but used his naked eye to compile all sorts of useful information about what he saw in the sky, ranging from a description of how the planets move to a detailed catalog of 1,022 stars. The *Almagest* remained the standard work of astronomy in Europe and the Middle East for the next 800 years.

As you've perhaps guessed, the modern-day *Almagest* is the Sloan Digital Sky Survey (SDSS), named for the Alfred P. Sloan Foundation, which provides much of the funding. The SDSS does its work using a superb telescope located just outside the tiny town of Sunspot, high in New Mexico's Sacramento Mountains. The telescope captures light using a large mirror, 2.5 meters in diameter. The excellent location and large mirror mean the SDSS takes very good images, and can look all the way out to the edge of the known universe. The images aren't quite as good as those from the world's biggest telescopes, such as the enormous 10.4-meter Gran Telescopio Canarias, in the Canary Islands. But the SDSS telescope has a major advantage over most larger telescopes: it has a special wide-angle lens that lets it rapidly photograph large sections of the sky. In a single image it can capture an area eight times the size of the full moon. By contrast, the Gran Telescopio Canarias can only capture an area one sixteenth the size of the moon, making it unsuitable for the broad exploration required by a sky survey. Since beginning operation in 2000, the SDSS has surveyed more than a quarter of the sky, taking images of 930,000 galaxies along the way. And, as we'll see, those images have since been used in thousands of other scientific projects, including the project of Gott and collaborators to make a map of the universe.

How do you go about making a map of the universe? It's a surprisingly complicated problem. Ideally we'd like a map to show both objects that are relatively close in astronomical terms, such as the nearby stars, which are just a few light-years away, and also the most distant galaxies, which are billions of light-years away. It's hard to do both those things on the same map. The mapmaking problem is also complicated by the fact that the universe is three-dimensional, while ordinary maps are two-dimensional. Of course, there are many ways you can try to address these complications, but that leads to still another problem: of the many ways you can make your map, which way is the best? A feature that's strikingly obvious in one way of visualizing the universe may be nearly invisible in another. And what if you make the wrong choice? Mapping the Earth's surface is a much easier problem, yet early mapmakers still tried out many different projections to make sense of the Earth. Similarly, Gott and his collaborators experimented with many different ways of making their map. One of the maps they made took the galaxy data from the SDSS, and used it to visualize the distribution of galaxies in the universe. That map is shown in figure 6.1. It's not an ordinary map like a roadmap, and so it takes a bit of effort to understand, but it's worth reading through the caption in detail to understand what's being shown. The key point is the concentration of galaxies in the upper left-hand corner of the map, a concentration much denser than through the rest of the map. It was humanity's first ever glimpse of the Sloan Great Wall.

The Sloan Great Wall is just one of thousands of scientific discoveries made using the SDSS. To give you more of the flavor of the SDSS's impact, let me briefly describe two more of those discoveries. You perhaps already know that our Milky Way galaxy has two neighboring galaxies, the Large and Small Magellanic clouds. These are dwarf galaxies, with the larger of the two containing about 30 billion or so stars, compared to our Milky Way's hundreds of billions. If you've never been to the southern hemisphere, then you may never have seen the Magellanic clouds, for they're too far south in the sky to be visible from much of the northern hemisphere. But they are visible on a dark night in the southern hemisphere, where they show up as smudges in the sky. According to our best current understanding of galaxy formation, the Milky Way should have tens or hundreds of nearby dwarf galaxies. But prior to the SDSS only a few dwarf galaxies other than the Magellanic clouds had been discovered, and it was a puzzle where all the other missing dwarfs were. When the SDSS images became available, several astronomers

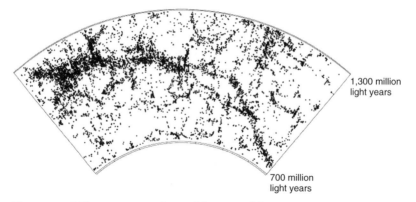

Figure 6.1. A blown-up piece of one of the maps of the universe made by Gott and collaborators. You'll notice that the map resembles a piece of pie. You should imagine yourself on the Earth, right at the center of the pie, looking out at the universe. Each point on the map represents a single galaxy from the SDSS. The radial direction indicates the distance to the galaxy, with the closest galaxies in the plot about 700 million light-years away, and the furthest about 1,300 million light-years away (as marked on the right-hand side). All the galaxies shown in the plot are very close to the celestial equator, the great arc going across the sky, directly above the Earth's real equator, and circumnavigating the Earth. What you're seeing here, then, are the galaxies in a thin slice of the sky, all very near the celestial equator. The angular direction in the plot shows where along the celestial equator the galaxy is located. Galaxies on the left-hand side of the map are in the direction of the constellation of Virgo, galaxies in the middle are in the direction of Leo, and galaxies on the right are in the direction of Cancer. The dense chain of galaxies concentrated in the upper left is the Sloan Great Wall. Credit: Reproduced by permission of the American Astronomical Society.

searched the images for more dwarf galaxies. They didn't do this manually—it would have taken far too long to peruse all the images. Instead, they used computer algorithms to search out new dwarf galaxies in the SDSS images. What they've found so far is nine new dwarf galaxies near the Milky Way, going much of the way toward solving the puzzle of the missing dwarfs.

As another example of an SDSS-enabled discovery, in 2009 the astronomers Todd Boroson and Tod Lauer used the SDSS to discover two black holes orbiting around one another. The way Boroson and Lauer found the paired black holes was—you won't be surprised!—by using a computer to search galaxy images from the SDSS. Now, black holes have no color, and don't show up directly in photos. But black holes are surrounded by huge amounts of glowing matter that's falling in, and so in a sense it's possible to "see" black holes in the galaxy images, to talk about them having a color, and so on. The key to Boroson and Lauer's work was a clever guess they made, which was that if two black holes were orbiting one another, they would appear to have slightly different colors. The reason they made this guess is interesting. When objects are moving at a high enough speed—it needs to be a considerable fraction of the speed of light—their apparent color changes appreciably. Why this happens is a long story, which we won't get into, but as an example, a red object that's moving very quickly toward the Earth actually looks a little bit bluer. Boroson and Lauer reasoned that two black holes orbiting one another would have different velocities relative to the Earth, and so one would be ever so slightly bluer than the other. Armed with this double coloring idea, Boroson and Lauer used their computer to go hunting in the SDSS data. Their hope paid off when they found a galaxy four billion light-years away with exactly the double coloring signature they were looking for. They followed up with a more detailed examination of the galaxy, confirming the presence of the orbiting black holes, and revealing that they are both staggeringly large, 20 million and 800 million times the mass of the sun, respectively, and a third of a light-year apart, orbiting one another roughly once every 100 years. The discovery has excited great interest, and also set off a debate, with other astronomers wondering if there might be some other explanation for what Boroson and Lauer are seeing. At the moment, the orbiting black hole theory remains the leading candidate among several possible explanations. But no matter what the truth turns out to be, no one doubts that Boroson and Lauer have discovered something remarkable.

All these discoveries are striking, but they don't fully convey the enormous impact the SDSS has had on astronomy. One way to grasp that impact is to look at how many times the results of the SDSS have been cited (i.e., referred to) in other scientific papers. Most papers in astronomy are cited just a few times, if they're cited at all. A paper that's cited tens of times is quite successful, while a paper that's cited hundreds of times is either famous or well on its way. The original SDSS paper has been cited in other papers more than 3,000 times. That's more citations than many highly successful scientists receive over their entire career. To give you some feeling for what an achievement this is, Stephen Hawking, probably the world's most famous scientist, has just a single paper with more than 3,000 citations. Hawking's paper, which he published in 1975, in fact has just over 4,000 citations as of 2011. By contrast, the SDSS paper was published in 2000, and already has more than 3,000 citations. It will soon catch up to and surpass Hawking's paper. Several follow-up papers describing other aspects of the SDSS have also received more than 1,000 citations. When I compared the SDSS to Ptolemy's *Almagest* I wasn't joking. The SDSS is one of the most successful ventures in the entire history of astronomy, worthy of a place alongside the work of Ptolemy, Galileo, Newton, and the other all-time greats.

Open Data

What's made the SDSS such a success? We've already discussed some of the reasons: the SDSS has an excellent telescope and broad coverage of the sky. But those can't be the only reasons. In the 1940s and 1950s astronomers used the giant 5-meter Palomar telescope outside San Diego, California, to carry out the Palomar Observatory Sky Survey. But while the Palomar telescope is in some respects even better than the SDSS telescope, the Palomar survey had a much less dramatic impact on astronomy. Why is that the case? The main reason is that the Palomar survey produced bulky photographic plates, which are expensive to move around and to duplicate, and so could only be accessed by a few people. By contrast, the SDSS has used the internet to share its data with the entire worldwide community of astronomers. Since 2001, the SDSS has done seven major data releases, putting its images (and other data) up on the web where anyone can download them. If you want, you can go right now to the SDSS's online SkyServer, and download stunning images of distant galaxies. Anyone can do it, and the site is designed to be used not just by professional astronomers, but also by members of the general public. The tools on the site range from tours of the most beautiful sights in the sky, through to the ability to send sophisticated database queries that will return images with particular desired characteristics. The site even contains tutorials explaining how to do things like find asteroids, or star-forming regions in other galaxies.

This open sharing of data by the SDSS seems like a small innovation when compared to the radical approaches to collective intelligence we saw in examples such as the Polymath Project and Kasparov versus the World. But the impact of the open sharing of data by the SDSS is enormous. It means that people such as Todd Boroson and Tod Lauer—people who aren't members of the SDSS collaboration—can come along and ask fundamental questions that no one had ever thought to ask before: "Can we use the SDSS data to search for pairs of orbiting black holes?" In science, discoveries are often constrained by the limits to our knowledge. But experiments such as the SDSS produce such an extraordinary wealth of knowledge—more than 70 terabytes of data, far beyond the ability of any single human to comprehend—that they confound that expectation. Confronted by such a wealth of data, in many ways we are not so much knowledge-limited as we are question-limited. We're limited by our ability to ask the most ingenious and outrageous and creative questions. By opening up its data to the whole world, the SDSS has enabled people such as Boroson and Lauer to ask such questions, questions that might never have been asked if access to the data was limited. It's the same thing we saw in Swanson's discovery of the migraine-magnesium connection: Swanson used no facts that weren't already known, but by asking a new question of existing knowledge, he made a valuable discovery. It's a variation on the designed serendipity we saw in part 1. Instead of broadcasting a question to the world and hoping for an answer, projects such as the SDSS broadcast data to the world, in the belief that people will ask unanticipated questions that lead to new discoveries.

The SDSS's sharing of data isn't just important because of the discoveries it enables. It's also important because sharing data in this way, as simple and obvious as it might seem, in fact is a radically daring step for the scientists involved. Most scientists guard their data jealously. Their data is their raw record of experimental observations, and may lead to important new discoveries. It's their special edge over their colleagues and competitors. Unusual as it may be for them to openly reveal their data, it's even more unusual for them to encourage their colleagues to make independent analyses, and perhaps independent discoveries. You can grasp something of what's at stake by looking at some famous cases where data was partially revealed. For instance, earlier I mentioned Ptolemy's *Almagest*, one of the great scientific works of antiquity. But I should perhaps have put "Ptolemy" in quotes, because many historians of science—not all, but many—believe that Ptolemy plagiarized many of the star positions in his catalog from the astronomer Hipparchus, who had done his own sky survey nearly 300 years earlier. In fact, the history of science is full of examples of scientists stealing data from one another. Back at the dawn of modern science the astronomer Johannes Kepler discovered that planets move in ellipses around the sun using data he stole from his deceased mentor, the astronomer Tycho Brahe. James Watson and Francis Crick discovered the structure of DNA with the aid of data they borrowed from one of the world's leading crystallographers, Rosalind Franklin. I say borrowed, because this was done without her knowledge, although with the aid of a colleague of Franklin's who was arguably within his rights. These are, admittedly, extreme examples, but they do show why most scientists go to some trouble to keep their data secret.

There's a puzzle here, then: why does the SDSS share data so openly? Think about the situation from the point of view of members of the SDSS collaboration. Almost certainly there are important discoveries that they could have made, but which they were beaten to by someone outside the collaboration who used SDSS data. To put it in starkly self-interested terms, while open data may be good for science, it's arguably bad for the careers of members of the SDSS collaboration. Why do they stand for it? Why doesn't the SDSS lock up the data?

In fact, the SDSS does partially lock up the data. When the SDSS telescope takes images, they aren't immediately made public. Instead, for a brief period of time—typically a few months to a little over a year—they are only available to official members of the SDSS collaboration. It's only after that period has elapsed that the data are made freely available to everyone in the world.

There's a similar partial openness about the membership of the SDSS collaboration. While most scientific experiments still involve only a small number of participants, the SDSS collaboration has 25 participating academic institutions, and includes also 14 additional scientists who are not at any of the participating institutions. All in all, roughly 200 scientists are official members of the collaboration, far more than was scientifically necessary to get the SDSS up and running. The home page of the website for the current phase of the SDSS (stage III) even encourages "[i]nquiries from interested parties to join the collaboration." Astronomy is a small community, with just a few thousand professional astronomers in the world. As a result many, perhaps most, professional astronomers have a friend or colleague who is part of the SDSS collaboration, and with whom they can potentially collaborate using SDSS data, even during the initial period when the data are not open.

These explanations clarify the process the SDSS uses to share data, but they don't answer our starting question, which is why the SDSS makes its data partially open in the first place. Why not just lock the data up for good? And why isn't the SDSS collaboration deliberately kept as small as possible, to increase the benefits received by individual members? Before I answer these questions, I want to briefly describe several more examples of experiments that make their data openly available. Those examples will help us understand why and when scientists make their data openly available, and why open data is important.

Building the Scientific Information Commons

In September of 2009 an organization called the Ocean Observatories Initiative began building a high-speed network for data and electricity on the floor of the Pacific Ocean. They're extending the internet to the ocean floor, with the eventual plan being to lay 1,200 kilometers (750 miles) of cable, from the shores of Oregon all the way up to British Columbia. This underwater internet will range more than 100 kilometers (60 miles) offshore. When it's complete, all manner of devices will be plugged into the network, from cameras to robot vehicles to genome-sequencing equipment. Imagine a volcano erupting underwater, and nearby genome-sequencing equipment switching on to take genetic snapshots of never-before-seen microbes vented during the eruption. Or imagine a network of thermometers and other sensors mapping out the underwater climate, much the way the SDSS is mapping out the universe. But the Ocean Observatories Initiative is going even further than the SDSS, making their data openly available right from the start, so anyone in the world can immediately download the data, looking for new patterns and asking new questions. What new discoveries will be made with this unprecedented knowledge of the ocean floor?

It's not just the oceans and the universe that are being mapped out. Efforts are now underway to build a map of the human brain. For example, scientists at the Seattle-based Allen Institute for Brain Science are building the Allen Brain Atlas, mapping out the brain down near the level of single cells, and determining which genes are turned on in which regions of the brain. It's an important step along the way to understanding how genes make a mind, and has the potential to be tremendously useful in understanding how our minds work. Scientists at the Allen Institute have sliced up 15 brains into hundreds of thousands of slices, each slice just a few microns thick. They then analyze each slice, determining which genes are turned on, and where. It's all done by a team of five robots that work around the clock, each robot analyzing 192 brain slices per day, every day. The Allen Institute scientists expect to complete their map of the brain by 2012, when the results will be made available as open data, for anyone in the world to download and analyze. An earlier effort by the Allen Institute, completed in 2007, has already given us an openly accessible map of how genes are expressed in the mouse brain. Furthermore, this work by the Allen Institute is part of a larger movement in neuroscience, toward an even more ambitious goal, mapping out the entire human connectome—the position of every neuron, every dendrite, every axon, and every synapse in the brain. It's possible that, one day in the not-too-distant future, we'll have a detailed, publicly accessible model of the entire human brain.

What we see in examples such as the SDSS, the Ocean Observatories Initiative, and the Allen Brain Atlas is the emergence of a new pattern of discovery. The SDSS is mapping out the entire universe. The Ocean Observatories Initiative will make broad-ranging observations of the ocean floor. The Allen Brain Atlas is mapping out the human brain. Still other projects aim to build detailed maps of the Earth's atmosphere, of the Earth's surface, of the Earth's climate, of human language, of the genetic makeup of all species. For just about any complex phenomenon in nature, chances are there's a project afoot to map out that phenomenon in detail. In many cases, it's not just a single project, but a whole pipeline of projects providing increasingly more detailed knowledge. We've seen this with human genetics, where the Human Genome Project mapped out the basic human genetic template; it was followed by the haplotype map, which mapped out the variations in human genetics; today, follow-up projects are getting still more detailed information about genetic variations in specific human groups. In astronomy, the SDSS will soon be succeeded by the Large Synoptic Survey Telescope (LSST), which will carry out a survey superior to the SDSS in nearly every way. The LSST, which is being built in the Andes of Chile, will be one of the world's largest telescopes, with an effective mirror diameter of 6.68 meters, much larger than the SDSS mirror, and so producing much better images. The telescope will have such an enormous field of view that it will map out the entire visible sky once every four days, instead of taking years to map out a fraction of the sky. Again, all the data will immediately be made openly available online.

Taken together, these and other similar projects are mapping out our world in incredible, unprecedented detail. Of course, similar survey projects have been undertaken through the whole history of science, from the *Almagest* to the great botanists of the eighteenth and nineteenth centuries. But what's going on today is special and unprecedented. The internet has dramatically expanded our ability to share and extract meaning from the models we are building. This has caused a corresponding increase in their scientific impact, as the SDSS vividly illustrates. The result is an explosion in the number and ambition of these efforts, bringing about a great age of discovery, much like the age of the explorers of the fifteenth to eighteenth centuries. But whereas those explorers went to the limits of the Earth's geography, the new discoverers are exploring and mapping out the boundaries of our scientific world.

As more data is shared online, the traditional relationship between making observations and analyzing data is changing. Historically, observation and analysis have been yoked together: the person who did the experiment was also the person who analyzed the data.

But today it's becoming more and more common for the most valuable analyses to be done by people outside the original laboratory. In some parts of science the division of labor is changing, with some people specializing in building the experimental apparatus and collecting data, while others specialize in analyzing the data from those experiments. In biology, for example, a new breed of biologist has emerged, the bioinformatician, whose chief skill isn't growing cell cultures or the other traditional skills of the biology lab, but who rather combines the skills of computer programmer and biologist to analyze existing biological data. In a similar way, chemistry has seen the emergence of cheminformatics, and astronomy the emergence of astroinformatics. These are disciplines where the main emphasis isn't on doing new experiments, but rather on finding new meaning in existing data.

Why is Data Being Made Open?

Let's return to the puzzle of why and when scientists make their data openly available. A clue comes from the size of the experiments. The SDSS, the Ocean Observatories Initiative, and the Allen Brain Atlas all cost (or will cost) tens or hundreds of millions of dollars, and involve hundreds or thousands of people. Our earlier examples of open data, such as the Human Genome Project and the haplotype map, were also enormous projects. But most scientific experiments are far smaller. And in the smaller experiments, open data is the exception, not the rule. Before I became interested in open data, I worked for 13 years as a physicist. In that time, I saw hundreds of experiments, nearly all of them small experiments done in modest laboratories. So far as I know, not a single one of those experiments made any systematic effort to make their data open. We saw something similar in the opening chapter, in the early reluctance of scientists to share genetic data in online databases such as GenBank. This has only changed because of major cooperative efforts such as the Bermuda Agreement on sharing human genetic data. Across science, the situation today is changing, with some scientific journals and grant agencies enacting policies that encourage or mandate that data be made openly available after experiments have been published. But open data remains the exception, not the rule. If you head out to your local university and walk into a small laboratory, you'll most likely find that the data is kept under lock and key, sometimes literally.

It seems, then, that big scientific projects are more likely to make their data open than small projects. Why is that the case? Part of the explanation is political. Think about the SDSS. A typical small astronomy project may cost "only" a few tens or hundreds of thousands of dollars. That's a lot of money, but it's small change out of the billions of dollars our society spends on astronomy. If the people doing the experiment keep the data to themselves, it's not a big loss to other astronomers. Furthermore, those other astronomers aren't in any position to complain, for they too are keeping the data from their experiments secret. It's a stable, uncooperative state of affairs. But the SDSS's size makes it special and different. It's so large that it consumes much of the entire world budget for astronomy. If the data is kept secret, then to astronomers outside the SDSS collaboration it's as though that entire chunk of money has simply disappeared from the astronomy budget. They have every reason to insist that the data be made open. And so, if large projects don't commit to at least partial openness, their applications for funding risk being shot down by people in the same field but outside the collaboration. This motivates big scientific projects to make their data at least partially open.

There is another factor inhibiting open scientific data, which is that even if you are willing to share your data, it can be difficult to do so in a way that's useful to others. You can take all the photographs of galaxies you like, and share them with others, but those photographs are of limited scientific use without all sorts of extra information. What color filters did you use? Has the image been processed in any way, say, to remove bad or damaged pixels? Was there any haze the night the photos were taken, which might obscure the image? And so on. In many parts of science it's difficult to make sense of experimental data without detailed calibration information. And even with the data *and* the calibration information, other scientists still need an extremely detailed understanding of the experiment to make use of the data. Add on top of that problems like being sure everyone is using technical terminology in exactly the same way, file format conversion, and so on. Individually these are all soluble problems, but together they're a formidable obstacle to sharing data in a way that's useful.

These questions about sharing data are part of a deeper story, a story about why and when scientific knowledge is shared. Earlier in the book, I mentioned several times that scientists build their reputation and career based on the papers they've written. A reputation for writing great papers will get them a good scientific job, and continued grant support. Much of the challenge with data sharing is that the rewards scientists get for sharing their data are much more uncertain than the rewards for writing papers. It's true that a few large collaborations such as the SDSS have won widespread kudos for sharing data. But in many areas of science, there are few established norms for how and when the use of someone else's data should be acknowledged. And that means that sharing data is chancy for a scientist. It's just not something scientists are typically well rewarded for, despite the fact that it's enormously valuable. And so open data remains uncommon, especially in smaller laboratories. We will return to the question of how to get scientists enthused about sharing data (and other related questions) in chapters 8 and 9. For the purposes of the remainder of this chapter it's enough that there is already a considerable (and increasing) amount of scientific data openly available, through projects such as the SDSS and the Human Genome Project.

Dreaming of the Data Web

So far in this chapter we've taken a concrete, near-term perspective, looking at existing projects such as the SDSS. But the internet is an infinitely flexible and extensible platform for manipulating human knowledge, with a potential that is open-ended. To understand that potential we need to expand our thinking, and move to a long view that sees the internet not as a ten- or twenty-year revolution, but as a hundred- or thousand-year revolution. We need to imagine a world where the construction of the scientific information commons has come to fruition. This is a world where all scientific knowledge has been made available online, and is expressed in a way that can be understood by computers. Imagine, furthermore, that the data aren't isolated in tiny little islands of knowledge, as they are today, with separate, siloed descriptions of phenomena that are fundamentally connected in nature, phenomena such as amino acids, genes, proteins, drugs, and human medical records. Instead, we'll have a linked web of data that connects all parts of knowledge. Rather than mining that knowledge in a piecemeal way, we'll be able to do automated inference on all of human knowledge, finding hidden connections on a scale that dwarfs the work of Swanson or even the SDSS. We'll give this dream a name: we'll call it the dream of the data web.

The data web sounds grandiose. But, as we've seen, we've already taken many small steps toward the data web, through projects such as the SDSS and the Human Genome Project. What's gradually emerging is an online network of knowledge that's intended to be read by machines, not by humans. Those machines will find meaning in that network of knowledge, and help explain it to us. In the remainder of this chapter we'll ask how the data web will be built, and what it will mean.

There is, however, a difficulty in the discussion, a difficulty that bedevils every discussion of the potential of computers to find meaning in knowledge: the more you speculate on this potential, the further you go in the direction of a discussion of full-blown artificial intelligence, the science-fictional the-internet-wakes-up-to-take-over-the-world type scenario. That's a lot of fun to talk about, but it's too easy to get bogged down in speculative questions: "So, can machines ever become conscious, and what is consciousness, anyway?" or "Well, yes, maybe one day the internet will wake up and take over, and what of it?" This is all ground that's been trodden many times before. Instead of repeating those discussions, we'll explore a middle ground between the near-term projects discussed earlier in the chapter, and full-blown artificial intelligence. This middle-ground future is conceptually rich, fascinating, and strangely under-discussed, perhaps because the dreams of artificial intelligence exert such a strong pull on the imaginations of the technologically curious. What we'll do is synthesize current ideas from computer science to understand what happens when you take today's algorithms and imagine a future in which they can be applied across all scientific knowledge. As we'll see, the likely results are spectacular.

Data-Driven Intelligence

To understand what the data web can be used to do, it helps to give a name to the ability of computers to extract meaning from data. I will call that ability *data-driven intelligence*. Examples of data-driven intelligence include the algorithms used in the Medline searches Don Swanson did to discover the migraine-magnesium connection, the algorithms used to correlate Google searches with CDC flu data, and the algorithms used to mine the SDSS for dwarf galaxies and orbiting black holes, and to discover the Sloan Great Wall.

The term "data-driven intelligence" is not new. But at present it is mostly used in a more restricted sense than what I'm proposing, to describe data-driven approaches to making corporate business decisions—for instance, the way airlines mine data on passenger no-shows to know how much to overbook their flights. I'm proposing to use the term in a much more general way, as a broad category of intelligence, similar to the way we use terms such as "human intelligence" and "artificial intelligence." In this general sense, "data-driven intelligence" is a much-needed term, partly because of the large and rapidly growing number of examples of data-driven intelligence. But what's even more important is that the term highlights a particular approach to finding meaning, an approach for which computers are superbly well suited, and which is different from and complementary to the way we humans find meaning.

Of course, a human chauvinist might object to my use of the term "intelligence" in "data-driven intelligence," arguing that there's nothing very intelligent about a computer searching ten million scientific papers, or searching the SDSS for dwarf galaxies. It's just routine, mechanical work, albeit on a scale far beyond human ability. But here's the point: these are problems we humans can't solve at all. When it comes to making meaning from terabytes or petabytes (thousands of terabytes) of data, we're not much better than any other animal. We have, at best, a few very specialized abilities in this domain, such as the ability to process visual images, and virtually no general-purpose large-scale data-processing ability. So who are we to judge computers in this domain? An unaided human's ability to process large data sets is comparable to a dog's ability to do

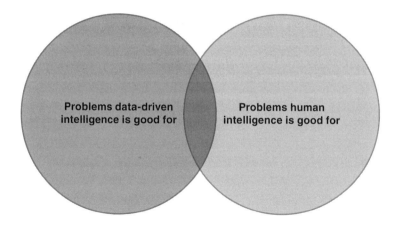

arithmetic, and not much more valuable. So while these problems perhaps don't require computers to be very smart, in this problem domain they are a lot smarter than humans. This point of view is captured in the diagram shown on this page.

It's human nature to focus on the problems on the right of the diagram, the problems where human skill and ingenuity are most valuable. And it's normal human prejudice to undervalue the problems on the left, the domain where data-driven intelligence really shines. But we'll put aside this prejudice, and think about the problems on the left. What problems can computers solve that we can't? And how, when we put that ability together with human intelligence, can we combine the two to do more than either is capable of alone?

As an example of the latter, in 2005 the chess website Playchess.com ran what they called a freestyle chess tournament, meaning a tournament where humans and computers could enter together as hybrid teams. To put it another way, the tournament allowed human intelligence to team up with data-driven intelligence, in the form of chess-playing computers, which rely on enormous opening and endgame databases, and which analyze myriad possible combinations of moves in the midgame. One of the entrants in the tournament was the team behind the Hydra series of chess computers. Hydra, at the time the world's strongest chess computer, had never lost a game in regular play to any human chess player, and on several occasions had easily defeated top grandmasters. The Hydra team entered two of their computers, one playing entirely on its own, and the other playing with some human assistance. Also entered in the tournament were several teams pairing strong grandmasters with strong chess computers. On their own, neither the grandmasters nor their computers could match the Hydras. But the joint human-computer teams trounced the Hydras. Not only did neither Hydra win the tournament, but in fact neither even made it to the quarterfinals. The grandmasters could beat the Hydras because they knew when to rely on their computers, and when to rely on their own judgment. Even more interesting, the winner of the tournament was a team called ZackS that consisted of two low-ranked amateur players using three off-the-shelf computers, and standard chess-playing software. Not only did they outclass the Hydras, they also outclassed the grandmasters with their strong chess-playing computers. The human operators of ZackS demonstrated exquisite skill in using the data-driven intelligence of their computer algorithms to amplify their chess-playing ability. As one of the observers of the tournament, Garry Kasparov, later remarked, "Weak human + machine + better process was superior to a strong computer alone and, more remarkably, superior to a strong human + machine + inferior process."

Data-driven intelligence has broader goals than artificial intelligence. For the most part, artificial intelligence takes tasks that human beings are good at and aims to mimic or better human performance. Think about computer programs to play human games like checkers, chess, and go, or efforts to train computers to understand human speech. Data-driven intelligence can be applied to these traditionally human tasks—it can understand human speech, or play chess—but where it really excels is in solving different kinds of problems, problems involving skills complementary to human intelligence, problems such as Swanson's searches of the medical research literature, or Boroson and Lauer's mining of the SDSS data for pairs of orbiting black holes. A full-fledged data-driven intelligence would be able to play checkers, chess, or go, but it wouldn't play them for fun. It would play games with a scope whose complexity was entirely beyond human comprehension.

The term "intelligence" is often used to mean some kind of generalized intellectual ability. Data-driven intelligence is more targeted in nature, with different kinds of data-driven intelligence used to solve different kinds of problems. We'll see an explicit example in the next section, which looks at the algorithms biologists use to do genome sequencing. A quite different set of algorithms is used to do searching in services such as Medline. For each problem, a different kind of data-driven intelligence is required. A consequence is that data-driven intelligence in some problem domain may start out quite stupid, but gradually get smarter as we develop improved methods. For instance, when Swanson did his migraine-magnesium work, search tools such as Medline used relatively simple ideas. Today's search engines use much more sophisticated ideas, and tomorrow's search engines will no doubt be much better still. Indeed, as data-driven intelligence helps companies such as Google turn a profit, those companies pour money into developing still better techniques, resulting in a virtuous circle of improvement.

How is data-driven intelligence related to collective intelligence? Actually, that's not quite the right question for our discussion. We're interested in data-driven intelligence as a way of augmenting our own intelligence, and so a better question is: how does data-driven intelligence relate to the tools we studied in part 1, the tools that *amplify* collective intelligence? As we saw, those tools work by restructuring expert attention so it's more effectively allocated. There's thus no direct relationship between tools that amplify collective intelligence and data-driven intelligence. But the two can be used in a complementary way. For instance, we've seen how data-driven tools such as Medline provide new ways of finding meaning hidden in the collective knowledge of large groups of people, such as the biomedical community. And data-driven tools such as Google can be used to amplify our collective intelligence by helping us find the information and the people that we should be paying attention to. Conversely, Google uses our collective intelligence to build its service, mining the web for content, and using the link structure of the web to figure out which pages are most important. So even though data-driven and collective intelligence are different, they can be used to reinforce each other.

This is not a textbook on data-driven intelligence, and I won't describe the hundreds of clever algorithms in use or under development. For us, data-driven intelligence is primarily important as a concept that unifies examples such as Google Flu Trends, the Sloan Great Wall, and Swanson's migraine-magnesium discovery. Underlying all these examples are clever algorithms that extract meaning from data that is otherwise beyond human ability to comprehend. Data-driven intelligence is in some sense complementary to the data web: although data-driven intelligence can be applied to any data source, it reaches its fullest potential when applied to the richest possible data sources, and the data web is the richest data source we can imagine. Data-driven intelligence is what will allow us to take all the world's knowledge and make meaning from it.

Data-Driven Intelligence in Biology

To make data-driven intelligence more concrete, let me describe in some detail an example from biology of how it works. The example shows how we can use clever algorithms and the scientific information commons to do something remarkable: find the genome of a human being. To understand the example, we first need to recall a little background about genetics. As you know, inside each of the cells in your body are many strands of the DNA molecule. Those strands of DNA carry information, and the information they carry is the design for you. To understand how DNA carries this information, recall the double helix structure of DNA. The helices are beautiful and memorable, but the information isn't stored in the helices, per se. Rather, it is stored in between the helices. Every few nanometers as you move up the double helix there is a pair of molecules joining the two sides of the helix, called a base pair. It's a pair of special little mini-molecules that bond to one another, and to the backbones of the double helix. There are four types of base molecule, called adenine, cytosine, guanine, and thymine. Their names are usually just shortened to A, C, G, and T. The A bonds to the T and the C to the G, so the possible pairs are A-T and C-G. You can thus describe the information in a single strand of DNA through a long sequence of letters—say, CGTCAAGG . . . — representing the bases bonded to one side of the helix (the other side will have complementary bases, GCAGTTCC . . .). That sequence is a description of your basic architecture. How exactly it specifies that architecture is still only partially understood, but everything we know suggests the sequence of DNA base pairs is the blueprint for our design.

How do we figure out the DNA sequence for a person? In fact, if we start with a fragment of DNA that is just a few hundred base pairs long, then it can be directly sequenced using clever old-school chemistry—essentially, one scientist, in their lab, carefully mixing chemicals. But if the DNA strand is much longer than that, then the problem of sequencing gets more complex. A typical strand of human DNA contains several hundred million base pairs, far too long to be sequenced directly. But there is a clever way of combining direct sequencing of short DNA strands with data-driven intelligence to figure out the full DNA sequence.

To understand how it works, imagine that I gave you a copy of the first Harry Potter book, *Harry Potter and the Philosopher's Stone*. But instead of giving you an ordinary copy of the book, I've taken a pair of scissors and cut the book into tiny little fragments. For example, the opening of the book might be cut up into these fragments:

"Mr. and Mrs. Dursley, of number four, Priv";

"et Drive, were proud to say that they we";

"re perfectly normal, thank you very much."

And so on. I've simplified things a bit here by showing the fragments in the same order they appear at the beginning of the book. But I want you to imagine that I've given them to you in the wrong order, all scrambled up. At the same time, imagine I have also given you a second copy of the book, also cut up into small fragments, but in a different way:

"Mr. and Mrs. Dursley, of num";

"ber four, Privet Drive, were proud to";

"say that they were perfectly normal, tha".

Even though the fragments in the two cases are different, there's quite a bit of overlap, and you can use those overlaps to figure out which fragments go together. Notice, for example, that the fragment "Mr. and Mrs. Dursley, of number four, Priv" overlaps with both "Mr. and Mrs. Dursley, of num" and "ber four, Privet Drive, were proud to." This suggests pasting the last two fragments together, to get "Mr. and Mrs. Dursley, of number four, Privet Drive, were proud to." By continuing very carefully in this way, you could reconstruct quite long sequences from the book. You'd only get stuck if, by chance, the overlap between two fragments was so short that it made it hard to tell that they really were overlapping fragments of the same text. But if I gave you a third (and a fourth . . .) copy of the book randomly cut up in this way, the chances of all the overlaps being short at any given point would drop dramatically, and you might well be able to reconstruct the entire book.

Genome sequencing for humans (and other complex life-forms) works in a similar way. While we can't directly sequence long strands of DNA, we can make many copies of those strands, then cut the copies up at random locations, and directly sequence the fragments. This can all be done using old-school chemistry, one scientist in their lab, etcetera. We then use our computers to figure out where different fragments overlap, and put everything back together again. (Incidentally, I've glossed over some subtleties, such as the repetition of certain DNA sequences throughout the human genome, which makes it harder to reassemble the full DNA sequence. These subtleties can be addressed using other tricks, but you now get the general idea.)

Now, imagine that we want to sequence someone's DNA today. Perhaps it's for a paternity test. Or maybe it's as part of a criminal investigation. It doesn't matter what the reason is. It turns out that we can actually simplify the above procedure for DNA sequencing, using the facts that (1) a reference human genome is already known, and (2) thanks to the haplotype map, we know where in the genome people may differ, and where, it seems, we're always the same. To understand how the simplified process works, imagine now that you possess a *complete* copy of *Harry Potter and the Philosopher's Stone*. Then, you're given a cut-up copy of a book that's similar, but that has been modified in a few locations. In fact, in real life the book really was changed between its initial release in the United Kingdom and its release in the United States. One change especially stands out, which is that the word *Philosopher* in the title was changed to *Sorcerer*, so the title became *Harry Potter and the* Sorcerer's *Stone*. All through the book "philosopher" was replaced by "sorcerer"— presumably, the publisher believed the book would have greater appeal in the United States this way. It's pretty obvious that having the complete text of the original book to refer to would make it much easier to figure out the text of the modified book. Instead of having to laboriously figure out which fragments matched with which, you could always figure out what part of the book the fragment you're currently examining is from. In a similar way, the sequencing of a human genome can be done faster and more easily by constantly referring back to the reference genome and the haplotype map.

Incidentally, while the Harry Potter example is fanciful, I can't resist mentioning that a very similar technique really *was* used by the author Chuck Hansen to write his book *U.S. Nuclear Weapons: The Secret History*. Hansen based his history on tens of thousands of declassified documents that had been sanitized by physically cutting out classified information. He discovered that different copies of the same document were sometimes sanitized in different ways, and by comparing different versions he could sometimes reconstruct the deleted information!

The algorithms I've described for genome sequencing are good examples of data-driven intelligence. In no sense are these algorithms especially smart. They're not doing much beyond simple pattern matching and rearrangement. But by combining these simple algorithms with enormous data processing power we can solve a problem that an unaided human being can't solve at all. Furthermore, by combining data-driven intelligence with the open data in the human genome and the HapMap we can simplify the problem of genetic sequencing. This is the kind of thing we'll see on a much grander scale when data-driven intelligence is combined with the data web.

Building the Data Web

Today, the data web is in its very early days. Most data is still locked up. To the extent data is shared, many different technologies are being used to do the sharing. The open data sets that are available mostly remain unconnected to one another, still living inside their separate silos. In short, the current state of the data web is messy and chaotic and incomplete. That's okay: the early days of a new technology are often messy. Think of how messy and chaotic the early history of aviation was, in the 1890s and early 1900s, before the Wright brothers first flew. Dozens of people were pursuing their own ideas about the best way to build heavier-than-air flying machines. It was out of that mess of ideas that the first airplanes slowly emerged. In a similar way, today thousands of people and organizations have their own ideas about the best way to build the data web. All are aiming in roughly the same direction, but there are many differences in the details. Perhaps the best-known effort comes from academia, where many researchers are developing an approach called the semantic web. In the business world, the state of affairs is more fluid, as companies try out many different ways of sharing data. Because of these many approaches, there are passionate arguments about the best way to build the data web, often carried out with great conviction and certainty. But the data web is still in its infancy, and it's too early to say which approach will succeed. For these reasons, I'll use the term "data web" rather loosely to refer to *all* open data, taken together in aggregate. It's a bit of an exaggeration, since much of that data isn't properly linked up, or is hard to find online. But that linking is coming, and so I've taken some license.

If we don't know what technology will ultimately be used to build the data web, how can we be sure the data web will grow and flourish? We can because a large and growing number of people want to share their data, and to link it up with other sources. We've seen a little of how this is happening in science. It's also true of many businesses and governments, some of which are making at least some data open. The website Twitter, for example, makes some of its data openly available, and this has led to the creation of third-party services such as TwitPic, which makes it easy to share photos on Twitter, and Tweetdeck, which offers a streamlined way of using Twitter. As another example, the day after US President Barack Obama's inauguration he issued a memorandum on "Transparency and Open Government." This memorandum led to the creation of a website called data.gov, where the US government shares more than 1,200 open data sets on subjects ranging from energy use to aviation accidents to television licenses. Examples such as these are driving the development of technologies to share data across the greatest number of users. Whichever technology wins broad adoption will become, by default, the data web. That's why we don't need to know which technological vision of the data web will win to conclude that the data web is inevitable.

Perhaps the most impressive steps toward the data web to date have been taken in biology. Biologists are picking off chunks of the biological world and mapping them out, building toward a unified map of all of biology. We've discussed some of these chunks— the human genome, the haplotype map, and the just-beginning human connectome. But there are many more. There are online databases that describe the biological world at a very small level, for example mapping out protein structure and function, and the many possible interactions between proteins (the "interactome"). There are online databases describing the large-scale biological world, mapping out things such as animal migration patterns, and even catalogs that attempt to map out all the world's species. And there are online databases at every level in between, a plethora of resources for the description of the biological world. Wikipedia's list of biological databases has more than 100 entries as of April 2011. Those databases can potentially be linked up, to reflect the connections in biological systems: genetic information is linked to protein information, which is linked to information about protein interactions, which is linked to information about metabolism, and so on, all building toward a unified map of biology.

Services are being developed to mine this nascent biological data web, sort of a Google-for-biology, able to quickly answer complex questions about life. Imagine a world of the future where the biological part of the data web has flourished. Imagine having the genome of newborn children immediately sequenced, and then correlated with a giant database of public health records to determine not just what diseases they're especially vulnerable to—an old trope of science fiction—but also what environmental factors might influence their susceptibilities to disease. "Your son has an 80 percent chance of developing heart disease in his 40s if he's sedentary in his 20s and 30s. But with three hours of moderate exercise each week that probability drops to 15 percent." As problems manifest, special drugs can be created, with their design tailored specifically to individual genetic makeup and past medical history.

Today, the biological data web is just a prototype. Life has tremendous complexity at many different levels, and we are just beginning to map out the biological world. Just settling the basic conceptual categories is challenging. Take the notion of a gene. Until recently, students were taught that a gene is part of the DNA that codes for a protein. That seems simple enough. But, in fact, what scientists mean by a gene is changing, as we come to better understand the relationship between DNA and proteins. The early insight that genes code for proteins is incomplete. We now know that the same sequence of DNA can sometimes be transcribed in different ways, into different proteins. At the same time, a single protein may be formed by transcribing DNA from several disconnected parts of the genome, sometimes even from genetic material on different chromosomes. These are just two of the many ways in which our notion of genes is currently changing. More generally, as our understanding of biology improves, many fundamental concepts are being redefined. And when that kind of redefinition happens it can have profound implications for the way we represent knowledge. It's easy to imagine at some point in the future a need to radically restructure our databases of knowledge, as we learn that our old conceptual schemas are wrong, and must be updated.

What the Data Web Will Mean for Science

As the data web flourishes, it will transform science in two ways. The first way will be to dramatically increase the number and variety of scientific questions that we can answer. We've already seen how the SDSS has enabled thousands of new questions in astronomy to be answered. The more data sources available, and the more richly they're linked, the more dramatic the effect will be. Think of the way Google's search data and the CDC flu data were combined. With either data set alone it's difficult to answer the question "Where is the flu happening, right now?" But when you have both data sets, you can answer that question. The result has a magical, free-lunch quality: combine two data sets and not only can you answer all the questions originally answered by those data sets, you can also answer surprising new questions that emerge from relationships between them. As the data web grows, so too will the number and variety of questions that can be asked. In some sense, the questions you can answer are actually an emergent property of complex systems of knowledge: the number of questions you can answer grows much faster than your knowledge. And the data web aspires to contain all the world's knowledge.

The second way the data web will transform science is by changing the nature of explanation itself. Historically, in science we prize explanations that are simple. Many of our greatest theories have a rabbit-out-of-the-hat quality, explaining many apparently different phenomena through a single core idea. For example, Darwin's theory of evolution by natural selection has one simple idea at its core, yet it is an astonishingly powerful framework for understanding the evolution of life. As another example, Einstein's general theory of relativity has been beautifully summarized in a single sentence, by the physicist John Wheeler: "Spacetime tells matter how to move; matter tells spacetime how to curve." That simple idea, when expressed mathematically, explains all gravitational phenomena, from the flight of a thrown ball, to the motion of the planets, to the origin of the universe. It's a miracle of explanation, and many scientists (myself included) experience an epiphany when first we understand it.

But some phenomena don't have simple explanations. Think about the problem of translating Spanish into English. These languages contain a great deal of accidental complexity, as a result of all the contingencies in their historical genesis. To make high-quality translations we have no choice but to deal with all that complexity. In everyday life translators do this in part through a wealth of knowledge about the details of the languages, and in part through hard-to-describe intuition, built up over years of exposure to both languages. Any really precise explanation of how to translate from Spanish to English will necessarily be quite complex, and certainly won't have the simplicity of the theory of evolution or the general theory of relativity.

Until recently, the complexity of the scientific explanations we use was constrained by the limitations of our own minds. Today, this is changing, as we learn how to use computers to build and then work with extremely complex models. To explain the change, let me give an example from the field of machine language translation. Starting around 1950, researchers began building computerized systems whose aim was to automatically translate from one language to another. Unfortunately, the early systems weren't very good. They tried to do the translation using clever, relatively simple models based on the rules of grammar and other rules of language. This sounds like a good idea, but despite a lot of effort, it never worked very well. It turns out that human languages contain far too much complexity to be captured in such simple rules.

In the 1990s researchers in machine translation began trying a new and radically different approach. They threw out the conventional rules of grammar and language, and instead started their work by gathering an enormous corpus of texts and translations—think, say, of all the documents from the United Nations. Their idea was to use data-driven intelligence to analyze those documents *en masse*, trying to infer a model of translation. For instance, while analyzing the corpus the program might notice that Spanish sentences containing the word "hola" often have the word "hello" in their English translation. From this, the program would estimate a high probability that the word "hola" results in the word "hello" in the translated text, while the probability for English words unrelated to "hola" ("tiger," "couch," and "January," for example) would be much lower. The program would also examine the corpus to figure out how words moved around in the sentence, observing, for example, that "hola" and "hello" tend to be in the same parts of the sentence, while other words get moved around more. Repeating this for every pair of words in the Spanish and English languages, their program gradually built up a statistical model of translation—an immensely complex model, but nonetheless one that can be stored on a modern computer. I won't describe the models they used in complete detail here, but the hola-hello example gives you the flavor. Once they had analyzed the corpus and built up their statistical model, they used that model to translate new texts. To translate a Spanish sentence, the idea was to find the English sentence that, according to the model, had the highest probability. That high-probability sentence would be output as the translation.

Frankly, when I first heard about statistical machine translation I thought it didn't sound very promising. I was so surprised by the idea that I thought I must be misunderstanding something. Not only do these models have no understanding of the meaning of "hola" or "hello," they don't even understand the most basic things about language, such as the distinction between nouns and verbs. And, it turns out, my skepticism is justified: the approach doesn't work very well—if the starting corpus used to infer the model contains just a few million words. But if the corpus has billions of words, the approach starts to work very well indeed. Today, this is the way the best machine translation systems work. If you've ever done a Google search that returned a result in a foreign language, you'll notice that Google offers to "translate this page." These translations aren't done by human beings, or by special algorithms handcrafted with a detailed knowledge of the languages involved. Instead, Google uses an incredibly detailed statistical model of how to do translation. It's far from perfect, but today it's the best automated translation system around. Shortly after launching their translation service, Google easily won an international competition for English-Arabic and English-Chinese machines translations. What's truly remarkable is that no one on the Google Translate team spoke Chinese or Arabic. They didn't need to. The system could learn to translate by itself.

These translation models are in some sense theories or explanations of how to translate. But whereas Darwin's theory of evolution can be summed up in a few sentences, and Einstein's general theory of relativity can be expressed in a single equation, these theories of translation are expressed in models with billions of parameters. You might object that such a statistical model doesn't seem much like a conventional scientific explanation, and you'd be right: it's not an explanation in the conventional sense. But perhaps it should be considered instead as a new kind of explanation. Ordinarily, we judge explanations in part by their ability to predict new phenomena. In the case of translation, that means accurately translating never-before-seen sentences. And so far, at least, the statistical translation models do a better job of that than any conventional theory of language. It's telling that a model that doesn't even understand the noun-verb distinction can outperform our best linguistic models. At the least we should take seriously the idea that these statistical models express truths not found in more conventional explanations of language translation. Might it be that the statistical models contain more truth than our conventional theories of language, with their notions of verb, noun, and adjective, subjects and objects, and so on? Or perhaps the models contain a different kind of truth, in part complementary, and in part overlapping, with conventional theories of language? Maybe we could develop a better theory of language by combining the best insights from the conventional approach and the approach based on statistical modeling into a single, unified explanation? Unfortunately, we don't yet know how to make such unified theories. But it's stimulating to speculate that nouns and verbs, subjects and objects, and all the other paraphernalia of language are really emergent properties whose existence can be deduced from statistical models of language. Today, we don't yet know how to make such a deductive leap, but that doesn't mean it's not possible.

What status should we give to complex explanations of this type? As the data web is built, it will become easier and easier for people to construct such explanations, and we'll end up with statistical models of all kinds of complex phenomena. We'll need to learn how to look into complex models such as the language models and extract emergent concepts such as verbs and nouns. And we'll need to learn how to cope with the fact that sometimes those emergent concepts will only be approximate. We'll need, in short, to develop more and better tools for extracting meaning from these complex models.

With all that said, it still seems intuitive that simple explanations contain more truth than complex explanations. This prejudice against complex explanations in science is so ingrained that it even has a name: we call it Occam's razor. The idea is that if we have two alternate explanations for the same phenomenon, we should prefer the simpler explanation. This belief is also reflected in other ways. When we come up with a single, simple explanation that explains a wide variety of apparently disparate phenomena, we're inclined to think that it's true. We shout "Eureka," we've found it, when something that seemed complex turns out to have a simple explanation. Think of Newton's amazing discovery that his laws of gravitation explain both how objects fall on Earth, and also the motion of the planets around the sun. Before Newton's discovery, those phenomena seemed completely separate from one another: how remarkable that the same laws explain both!

Our confidence in the truth of simple explanations is so great that when we discover apparent violations of such an explanation, we may go to great lengths to save it. In the 1970s the astronomer Vera Rubin discovered that stars toward the outer reaches of our Milky Way galaxy are rotating around the center of the galaxy much faster than we'd expect on the basis of our best theory of gravity, the general theory of relativity. But rather than give up on general relativity, most astronomers instead prefer to postulate the existence of invisible dark matter permeating the galaxy. If the distribution of dark matter is just right, then general relativity can correctly account for the speed of stars on the outer edges of the galaxy. By comparison to the popularity of dark matter, new theories of gravity have been pursued by relatively few astronomers.

So far I've made little distinction between conventional explanations and complex models. This blithe conflation of the two has perhaps bothered some readers. Many people believe there is a hard and fast distinction between an explanation and a model: explanations contain some element of the truth, while models are merely convenient crutches, useful for illuminating some phenomenon, but ultimately not expressing the truth. This point of view has an intuitive appeal, but in the history of science the distinction between models and explanations is blurred to the point of nonexistence. Ideas that start out as "mere" models often contain the seed of truths that surprise even their originators. In 1900 the physicist Max Planck was trying to understand how the color and intensity of light emitted by an object depend upon its temperature. For example, burning coals at first glow red, but as the coals heat up, they change color and will eventually glow blue. Figuring out the relationship between temperature and color was a puzzle because the best physical theories of the day gave two different answers, both of which were contradicted by experiment! Planck tried many ideas to solve the problem, eventually settling on a model in which he made the ad hoc assumption that the energy associated with light must come in quantized packets, that is, must be a multiple of some basic unit. This was an arbitrary assumption, and Planck himself later said, "I really did not give it much thought"—it was just a trick that led him to the result he wanted. In fact, it turned out that the idea in Planck's model was ultimately the seed for one of the great discoveries of science, the theory of quantum mechanics. So should we regard Planck's ideas as merely a model, or as an explanation? At the time, it looked like a model, but that model contained a truth deeper than any of the theories of the day. In any reasonable accounting, Planck's ideas are both a model and an explanation: models and explanations are both part of the same continuum. And so, as online tools enhance our ability to construct and extract meaning from complex models, they will also change the nature of scientific explanation.

CHAPTER 7

Democratizing Science

On August 7, 2007, a 25-year-old Dutch schoolteacher named Hanny van Arkel was surfing the web when she came across the Galaxy Zoo website. As you may recall from the opening chapter, Galaxy Zoo recruits volunteers to help classify galaxy images. The volunteers are shown photographs of galaxies—often, galaxies no human has ever before seen—and asked to answer questions such as "Is this a spiral or an elliptical galaxy?" or "If this is a spiral, do the arms rotate clockwise or anticlockwise?" It's a kind of cosmological census, the largest ever undertaken, with more than 200,000 volunteers so far classifying more than 150 million galaxy images. When she came across Galaxy Zoo, van Arkel was immediately hooked, and she began classifying galaxies in her spare time. A few days after joining, she noticed a strange blue blob hovering just below one of the galaxies. What she saw is reproduced on the next page, in black and white, with an arrow pointing to the blob.

Puzzled, on August 13 she posted a note to Galaxy Zoo's online forum, asking if anyone knew what the blue blob might be.

No one knew.

Tests were done. The blob wasn't some kind of blemish in the photograph, it was real. Observations were made at other telescopes to get more detailed information, including observations with the powerful William Herschel telescope in the Canary Islands. Those observations showed that the blue blob was at about the same distance from the Earth as the galaxy hovering above it, which meant the blob was enormous, tens of thousands of light-years in diameter. More experts were called in, none of whom had ever seen anything like it.

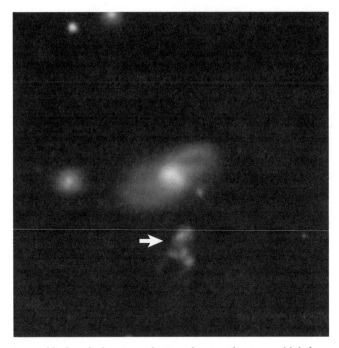

Figure 7.1. A black and white reproduction showing the strange blob first noticed by Hanny van Arkel. In the original color image the blob was a striking blue, and contrasted vividly with the galaxy above. Credit: Sloan Digital Sky Survey.

The mystery mounted. More and more people began speculating about what the blue blob could be. The object was dubbed Hanny's Voorwerp, after the discoverer and the Dutch word for object.

Slowly, an explanation of the voorwerp took shape, an explanation connecting the voorwerp to the staggeringly bright objects known as quasars. To understand that explanation, we first need to back up a bit and talk about quasars. As you may know, quasars are among the strangest and most mysterious objects in the universe. They are incredibly bright: a quasar the size of our solar system can shine as brightly as a trillion suns, outshining a giant galaxy like our Milky Way many times over. Fortunately for us, the nearest quasars are hundreds of millions of light-years away—if a quasar turned on a few light-years away, it would fry the Earth.

When quasars were first discovered, in 1963, it was a mystery how such comparatively small objects could shine so brightly. It took astronomers and astrophysicists many years to understand and agree on what is going on, but by the 1980s it was widely accepted that quasars are powered by solar system–sized black holes at the center of galaxies. Those black holes devour surrounding matter—stars, dust, you name it—while other matter swirls around the black hole, not quite falling in, but accelerated to near the speed of light. That enormous acceleration produces vast quantities of energy, some of which is emitted as light. It's that light that we see on Earth as the quasar. But while this rudimentary picture of quasars is now widely accepted, many fundamental questions remain unanswered.

With that understanding of quasars in mind, let's come back to the voorwerp. As the people at Galaxy Zoo puzzled over what the voorwerp might be, they considered many possible explanations, and gradually closed in on a simple explanation that seemed to fit all the facts: the voorwerp is a quasar mirror. The idea is that about 100,000 years ago, the galaxy near the voorwerp contained a quasar. That quasar has since switched off, for reasons unknown, and we no longer see it. But while the quasar was still shining, the light from the quasar was heating up gas inside a nearby dwarf galaxy, and causing it to glow. It's that glowing gas that we now see as a blue blob, and that's why we can think of the voorwerp as a quasar mirror. In fact, it's really a huge collection of mirrors, distributed over a vast region of space, echoing the light of the quasar at many different times in its history. Of course, I'm using the term "mirror" loosely here, since the light from the voorwerp isn't reflected light, but is instead the glow of heated gas. It's a sort of light echo from the quasar.

Not all astronomers and astrophysicists find the quasar mirror explanation convincing. To some, it seems a bit too convenient that the quasar has switched off. Another group has put forward an alternative explanation for the voorwerp, involving a different type of source in the nearby galaxy, a source that is also black hole–powered, but that is not a quasar. This presumed source is called an active galactic nucleus (AGN). It's a super-massive black hole that's emitting what's called a jet, a narrow cone of plasma tens of thousands of light-years long. By chance the jet is aimed in the direction of the voorwerp. The jet heats up the gas in the voorwerp and causes it to glow. So in this explanation the voorwerp isn't a quasar mirror, it's an AGN mirror (again, loosely speaking)!

As I write, astronomers and astrophysicists are still trying to figure out which explanation is correct. But regardless of which explanation is correct, or even if some other explanation is needed, the voorwerp is fascinating. Suppose, for instance, that it really is a quasar mirror. As we've seen, this means that the voorwerp is a huge collection of mirrors, echoing the light of the quasar at many different times over the quasar's lifetime. That means light from the voorwerp is a bit like a biography of the quasar, and by examining the voorwerp very closely we may learn a lot: how the quasar lived, how it died, and maybe even how it was born. That makes the voorwerp tremendously important as a way of studying the life cycle of quasars. Similarly, if the voorwerp is really shining because of a jet from an AGN, studying the voorwerp will be a great way of learning more about AGNs. In either case, astronomers are excited by the possibilities, and plan follow-up investigations aimed at getting a more detailed picture of the voorwerp. Observation time has been obtained on some of the world's most in-demand telescopes, including the Hubble and other space-based telescopes. From these and other observations we will learn more about the voorwerp, and perhaps about quasars or active galactic nuclei, too. The story of the voorwerp is just beginning.

Redefining Science's Relationship to Society

We take it for granted that science is for the most part done by scientists. Part of what makes Hanny's Voorwerp exciting is that it violates this assumption. How remarkable that a 25-year-old schoolteacher has discovered this great and beautiful cloud of gas! How unexpected that an amateur could make a discovery that might change our understanding of quasars or active galactic nuclei! When the discovery of the voorwerp was announced, it was a media story all over the world, receiving coverage on CNN and the BBC, in *The Economist*, and in many other major media outlets. Although I was delighted for Hanny van Arkel and the Galaxy Zoo team, as a writer my first feeling about all this publicity was a certain selfish disappointment, thinking that I would need to remove the voorwerp from my book, and replace it with a fresher example.

But after more thought I decided to leave the voorwerp in: the media splash itself illustrates just how strongly we take it for granted that science is done by scientists, and how fascinated we are by exceptions to this rule. The headline at CNN says it all: "Armchair Astronomer Discovers Unique 'Cosmic Ghost.'" What a shock and surprise that a nonscientist could make a significant astrophysical discovery!

Galaxy Zoo and the voorwerp are part of a bigger story about how online tools are gradually changing the relationship between science and society. One of the most fertile areas where this is happening is citizen science, with projects such as Galaxy Zoo recruiting online volunteers to help make scientific discoveries. In the first half of this chapter, we'll look at citizen science in depth, seeing how it changes who can be a scientist, and how it enables new types of scientific problem to be attacked. But citizen science is not the only way online tools are changing the relationship between science and society. In the second half of the chapter, we'll look at other new bridging institutions enabled by online tools, and consider how such institutions may change the role of science in public debate and decision making. This discussion perhaps seems tangential to the main theme of the book, since it doesn't directly relate to how scientists make discoveries. But over the long run these social changes may greatly alter the context in which science is done, and it's worth exploring them in some depth. First, let's return to examine Galaxy Zoo in more detail.

Galaxy Zoo Revisited

> I can honestly say that Galaxy Zoo is the best thing I've ever done.... I don't know quite what it is, but Galaxy Zoo does something to people. The contributions, both creative and academic, that people have made to the forum are as stunning as the sight of any spiral, and never fail to move me.
> **—Alice Sheppard, volunteer Galaxy Zoo moderator**

Galaxy Zoo began in 2007, with two scientists at Oxford University, Kevin Schawinski and Chris Lintott. As part of his PhD work,

Schawinski was looking at photos of galaxies. Galaxies come in many shapes and sizes, but most galaxies are either spiral galaxies, like our own Milky Way, or else elliptical galaxies, roughly spherical balls of stars and gas. Conventional wisdom in 2007 held that most of the stars in elliptical galaxies are very old stars, getting up toward 10 billion years in age. When stars get old, they will often change color and size, turn into red giants, with the result that many elliptical galaxies have a reddish tinge when compared with spiral galaxies, which are younger, and contain many newly formed blue stars.

Schawinski suspected that the conventional wisdom was wrong, that some elliptical galaxies might not be so old after all, and there might be a lot of star formation going on inside them. To test his suspicion, Schawinski spent a week poring over photos of 50,000 galaxies from the Sloan Digital Sky Survey (SDSS), looking to see which of the galaxies were elliptical and which were spiral. As I mentioned in the opening chapter, distinguishing elliptical and spiral galaxies is something humans still do better than computers. Once he finished the classification Schawinski used a computer program to analyze each elliptical galaxy, to see how red or blue it was. As he had suspected, the results suggested that the conventional wisdom was wrong, that star formation *was* going on in some ellipticals. Unfortunately, the effect was weak, and he needed to analyze a much larger sample of galaxies to really nail it down. Fortunately, as we discussed in the last chapter, the SDSS had made images of 930,000 galaxies openly available. This was a promising but daunting resource. Classifying the first 50,000 galaxies had involved a heroic weeklong effort by Schawinski—to classify 50,000 galaxies over seven 12-hour working days requires classifying an image every six seconds! Even at that tremendous pace, it would take many months to classify 930,000 galaxies. And there's no way Schawinski could maintain that pace. Even if he devoted most of his working time to the classification, it would take years.

One day in March of 2007, Schawinski adjourned to the Royal Oak, a pub in Oxford, together with a postdoctoral scientist who had recently arrived at Oxford, Chris Lintott. Over a pint they considered a wild idea for classifying the SDSS photos. Instead of doing the classification work themselves, perhaps they could build a website that would invite members of the general public to help out. They dragooned some friends who worked as web developers to help build the site, and on July 11, 2007, the Galaxy Zoo site went live with an announcement on BBC Radio 4's *Today* program.

The response to the announcement of Galaxy Zoo dwarfed expectations, overwhelming and quickly crashing the new website. For the next six hours the Zookeepers running the site worked frantically to get the site back up and running. When the site finally reappeared, volunteers rapidly began signing up, and by the end of the first day more than 70,000 galaxy classifications were being done every *hour*—more than Schawinski had managed in his heroic week. Each galaxy was examined independently by many volunteers, enabling the Zookeepers to automatically identify and discard incorrect classifications. This made the results comparable to careful classification by professional astronomers. Although the rate of galaxy classification gradually slowed from its peak of 70,000 per hour, Galaxy Zoo's first classification of galaxies was complete after just a few months. That gave Schawinski the data he needed to finish his project. Verdict: yes, the conventional wisdom about spirals versus ellipticals was wrong, and some ellipticals really do contain a lot of newly formed stars.

Galaxy Zoo began with Schawinksi's questions, but over time the site has expanded to address a much broader range of questions. Many discoveries have been made serendipitously, when some Zooite (as the participants call themselves) has noticed something unusual in a photo, as in Hanny van Arkel's discovery of the voorwerp. A second, more complex example of serendipitous discovery is the story of the "green pea" galaxies. This story illustrates the potential of citizen science even better than the voorwerp, and so I'll recount it here. Incidentally, my account draws on a marvelous article written by one of the Zooites, Alice Sheppard, which you can find referenced in the "Notes" at the end of the book.

On July 28, 2007, two weeks after the Zoo first opened, a poster to the Galaxy Zoo forum named Nightblizzard posted a picture of a fuzzy green galaxy, noting that it was unusual for galaxies to be green. A couple of weeks later, on August 11, 2007, someone else posted a picture of a strange green galaxy. It was unusually bright, and the poster, named Pat, asked if the galaxy might be a quasar. No one was quite sure.

The next day, on August 12, a third poster, the ubiquitous Hanny van Arkel, found another of the strange green galaxies. Van Arkel dubbed the galaxy a "green pea," and posted it to the forum with a message titled "Give peas a chance!" Other Zooites thought this was hilarious, and started to dig up peas of their own, adding them to the "pea soup" taking shape on the forum. For several months the discussion thread grew. At first it was mostly people adding objects, or making pea jokes ("peas stop"). But people also asked thoughtful questions. What exactly were the peas? Why hadn't anyone heard of them before? One poster commented: "They talk about stars, galaxies, nebulae, planets, etc. in astronomy courses, but they never mention the peas. It must be a big secret among professional astronomers. They probably want all the peas for themselves to eat."

At first, pea collection was just a fun hobby for the Zooites. But as the collection of peas grew, so too did the mystery surrounding them. Some turned out to be ordinary stars or nebulae. But a few of the green galaxies still stood out as unusual. The Zooites figured out—I'll describe how shortly—that some of the pea galaxies were surrounded by incredibly hot, ionized oxygen gas. That was unusual for a galaxy. What were these small, green, highly luminous galaxies, surrounded by hot, ionized oxygen? And why had nobody ever heard of them before?

Let me pause here to explain how the Zooites figured out that the peas were surrounded by hot, ionized oxygen. It's an interesting piece of science, and illustrates just how serious some of the Zooites were becoming. Obviously, they couldn't determine that oxygen was present by going and visiting one of the galaxies. Instead, they figured it out by teaching themselves a technique called spectral analysis. We don't need to go into the details of how spectral analysis works, but the basic idea is quite simple. It's based on what's called the spectrum of a galaxy. What the spectrum shows is how the light from a galaxy breaks up into different colors—say, a little bit of red, a lot of green, and a dash of blue. In fact, the spectrum can even show (for example) that the light is a mixture of several slightly different shades of green, exactly what shades those are,

Figure 7.2. The first of the green pea galaxies, found by Galaxy Zoo forum member Nightblizzard in July, 2007. The green pea is in the center. Like all of the peas, it looks quite nondescript, and if you're not familiar with galaxies, it's tempting to think it's just another elliptical galaxy, or maybe a star. But many of the Zooites became quite expert at analyzing galaxy images, and it wasn't long before they realized the peas were unusual. Credit: Sloan Digital Sky Survey.

and their respective proportions. So the spectrum is a very detailed and precise way of breaking down galaxy images into their different colors.

The reason the spectrum of a galaxy is important is because it allows astronomers to figure out what the galaxy is made of. This may sound surprising, but again the idea is quite simple: when you heat up a material, say, sodium, it tends to glow with a particular mixture of colors. That's why sodium streetlamps glow with a very particular yellow-orange color. It turns out that every material—not just sodium, but oxygen, hydrogen, carbon, and any other—has its own unique spectrum, that is, glows with a characteristic mixture of colors. The spectrum of a material is thus a bit like a signature, and by looking closely for such signatures in a galaxy's spectrum it's possible to figure out what the galaxy is made of. It's one of the more remarkable discoveries of science: by looking carefully at the color of distant objects we can figure out what they're made of, and even how hot they are, since heating a material up changes its characteristic spectrum slightly. The SDSS made high-quality spectra available for all Galaxy Zoo galaxies, and it was by looking closely at the spectrum of the green peas that the Zooites figured out that some of the peas were surrounded by hot, ionized oxygen gas.

(I can't resist digressing to mention the marvelous fact that the substance helium was actually discovered using spectral analysis! In 1868, the astronomers Pierre Jules César Janssen and Joseph Norman Lockyer independently observed that the spectrum of the sun had features unlike any substance ever seen on Earth. They deduced, correctly, that they were seeing the first sign of a new chemical substance. But it wasn't until almost 30 years later that a chemist named William Ramsay discovered helium on Earth.)

Enough about spectral analysis; back to Galaxy Zoo and the mystery of the green peas. By this point—December 12, 2007—Zookeeper Kevin Schawinski had become intrigued by these strange galaxies. He decided to take a closer look at the peas. He ran some tests and quickly confirmed that they were indeed a new type of galaxy.

You might think the professional astronomers would now move in and take over the project. After all, the amateurs at Galaxy Zoo had just discovered an entirely new class of galaxy! But the pros, including Schawinski, were busy with other things, including Hanny's Voorwerp, and they didn't take over straight away. Instead, what happened next was a remarkable piece of science driven by the amateurs. The tone was set by a Zooite named Rick Nowell. Nowell went back through all the pea images that had been posted to the Galaxy Zoo forum, and systematically identified 39 objects that looked like they might be the new type of galaxy. Inspired by Nowell's list, other people started to make their own lists, and began debating what criteria should be used to distinguish this new type of galaxy from similar-looking objects, such as green stars. The tone of the project began to change, becoming focused on getting to the bottom of the pea mystery. People found red galaxies with characteristics similar to the green peas, but further away. More and more, the discussion focused on detailed properties of the galaxies' spectra, and several of the Zooites became quite adept at spectral analysis—the kind of expertise usually the province of professional astronomers.

The back-and-forth discussion of ideas at this stage was astonishing. I'd like to give a blow-by-blow account, but it would take far too long even to summarize here—this isn't a book about how to discover and understand a new type of galaxy! But what was especially remarkable about the discussion was its style. It's the kind of discussion any scientist recognizes. Scientific discoveries often begin with a bit of a mystery, vague suspicions, and some half-baked ideas—just like the initial vague suspicion that the green peas might be a new type of galaxy. That initial suspicion is gradually refined. New ideas are introduced, tested, improved, and sometimes discarded. Participants become obsessed, as their suspicions slowly turn into hard, detailed fact. This is the process of research, familiar to any research scientist, and it's exactly what you see in the Galaxy Zoo discussion of the green peas. It's eerily reminiscent of the discussions in the Polymath Project. The Zooites may be amateurs—they know far less about astronomy than many of the polymaths do about mathematics, and there is more levity in the Galaxy Zoo discussion—but underneath these differences, there is the same fertile sense of ideas growing and being refined, of a conviction that there is something here to be known, and a determination to get to the bottom of it. The Zooites don't have the credentials of some of the polymaths. But they are scientists.

As the Zooites gradually developed more precise criteria characterizing the green pea galaxies, they also became more sophisticated in how they found candidate images. No longer were they just sifting through Galaxy Zoo images by hand. Instead, they went to the original SDSS data, and developed sophisticated database queries that automatically searched the entire SDSS data set for galaxies that fit their criteria. Those candidates were then closely scrutinized by volunteers, and a list of 200 or so drawn up that seemed likely to be the new type of pea galaxy.

The professionals watched all this discussion with interest, and in early July of 2008 Schawinski, now a postdoctoral scientist at Yale University, and a Yale student named Carolin Cardamone decided to ramp up their involvement. In collaboration with the Zooites, Cardamone and Schawinski began detailed spectral analyses of the peas using sophisticated computer software. Over the next nine months they completed the work begun by the Zooites. The picture of the peas that emerged showed that they were, indeed, a new type of galaxy. They were ultra-compact, less than 10 percent the mass of our Milky Way galaxy, and forming stars very quickly—whereas the Milky Way produces just one or two new stars every year, the peas produce more like 40 new stars per year, despite being far smaller. And the galaxies were extremely bright for their size.

The green peas and the voorwerp are just two of the many discoveries made by Galaxy Zoo. Another Galaxy Zoo project was to search out images of merging galaxies (see the image on the next page). Mergers are life-changing events for galaxies, and so understanding mergers is of great interest to astronomers and astrophysicists. Our own Milky Way is currently merging with several small dwarf galaxies, and has been predicted to one day merge with the giant Andromeda galaxy, currently two million light-years away. Unfortunately, despite their importance, merging galaxies aren't so easy to find, and as a result most studies of mergers use samples containing only a few dozen merging galaxies. The Galaxy Zoo merger project quickly found 3,000 merging galaxies, a treasure chest of mergers for future studies. Other objects the Zooites have gone hunting for include gravitational lenses (objects whose gravity actually warps and focuses the light from objects that are farther away), and paired galaxies (galaxies that appear to be on top of one another, but where one galaxy is actually much closer than the other). There's even a voorwerp project, and the Zooites have successfully hunted down several more voorwerps.

In all, Galaxy Zoo has been used to write 22 scientific papers, on a wide variety of topics, and many more papers are on the way. The discoveries are sometimes serendipitous, as in the case of the voorwerp, and sometimes based on systematic analysis, as in the mergers project. Sometimes serendipity is followed up with extensive systematic analysis, as in the study of the green peas. Follow-up projects Galaxy Zoo 2 and Galaxy Zoo: Hubble have launched, and are providing even more detailed information about some of the galaxies observed by the SDSS, and also by

Figure 7.3. Two merging spiral galaxies (known jointly as UGC 8335). Credit: NASA, ESA, the Hubble Heritage (STScl/AURA)–ESA/Hubble Collaboration, and A. Evans (University of Virginia, Charlottesville/NRAO/Stony Brook University).

the Hubble Space Telescope. Other new projects from the team that started Galaxy Zoo include Moon Zoo, which aims to better understand the craters on the moon, and Project Solar Storm Watch, which aims to spot explosions on the sun. One of the astronomers involved in Galaxy Zoo 2, Bob Nichol of the University of Portsmouth, contrasted Galaxy Zoo with everyday astronomy in this way:

[In my everyday work] I can ask the question "how many galaxies have a bar through the middle of them" and typically I would embark on a career-long quest to answer this fundamental question. I may even recruit some poor graduate student to eyeball 50,000 galaxies to answer the question (like they did with Kevin!). But now, two days after the launch [of Galaxy Zoo 2], we already have the data to address this question and it's a little too fast for an old-timer like me.... The internet is clearly the revolutionary technology of this generation of astronomer.... Galaxy Zoo is an amazing demonstration of how powerful this new tool can be [when] used to address new questions.

Like a computer, Galaxy Zoo can find patterns in large data sets, data sets far beyond the comprehension of any single individual. But Galaxy Zoo can go beyond computers, because it can also apply human intelligence in the analysis, the kind of intelligence that recognizes that the voorwerp or a green pea galaxy is out of the ordinary, and deserves further investigation. Galaxy Zoo is thus a hybrid, able to do deep analyses of large data sets that are impossible in any other way. It's a new way of turning data into knowledge. Time and again, the Zookeepers meet new astronomers who say that their work could be aided by Galaxy Zoo, and more than twenty astronomers are now using Galaxy Zoo as a way of studying a broad range of astronomical questions. Galaxy Zoo is rapidly becoming a general-purpose platform connecting professional astronomers to interested members of the general public, so they can do science together.

When Amateurs Rival Professionals

It's not just in astronomy that citizen science is useful. One of the big open problems in biology is to understand how the genetic code gives rise to an organism's form. Of course, we've all heard many times that DNA is the "blueprint for life." But even though the slogan is familiar—it is, after all, the fate of great slogans to become cliches—that doesn't mean anyone yet understands in detail how DNA gives rise to life. Suppose biologists had never seen an elephant's trunk. Could they look into an elephant's DNA and somehow see the trunk there—that is, predict the trunk's existence based solely on the sequence of base pairs in an elephant's genetic code? Today, the answer to this question is no: how

DNA determines an organism's form is one of the mysteries of biology.

To help solve this mystery, a citizen science project called Foldit is recruiting online volunteers to play a computer game that challenges them to figure out how DNA gives rise to the molecules called proteins. That challenge may sound a far cry from deducing the existence of the elephant's trunk—it *is* a far cry—but it's a crucial step along the way, because proteins carry out many of the most important processes in our bodies. Aside from its intrinsic scientific interest, Foldit is also interesting as a demonstration of the great complexity of work that can be done by volunteers. In Galaxy Zoo, participants mostly carry out simple tasks, such as classifying a galaxy as spiral or elliptical. In Foldit, players are asked to tackle tasks that would challenge a biochemistry PhD. And, as we'll see, the top Foldit players are doing those tasks extraordinarily well.

Before we discuss Foldit in detail, let's talk a bit about proteins in general. Biologists are obsessed by proteins, and with good reason: they're molecules that do everything from digesting our food to contracting our muscles. A good example of a protein is the hemoglobin molecule. Hemoglobin is one of the main components in our blood: it's the molecule our bodies use to move oxygen from our lungs to the rest of our body. Another important class of proteins are the antibodies in our immune system. Each antibody has its own special shape that lets it lock on to viruses and other intruders in our body, tagging them for attack by our immune system.

At present we only partially understand how DNA gives rise to proteins such as hemoglobin. What we do know is that certain sections of our DNA are protein coding, meaning that they describe a specific protein. So, for example, there's a protein-coding section for hemoglobin somewhere in your DNA. That region is a long string of DNA bases, which starts: CACTCTTCTGGT.... It turns out to be helpful to divide that string of bases into triplets, which are called codons: CAC TCT TCT GGT.... The way proteins are formed is that each codon in the protein-coding section of your DNA is transcribed into a corresponding molecule in the protein called an amino acid. So, for example, the first codon for hemoglobin, CAC, gets transcribed into an amino acid known as histidine. I won't explain exactly what histidine is, or what it does—

for us it doesn't much matter. What matters is that everywhere the CAC codon appears in the DNA sequence for hemoglobin (or any other protein), it gets transcribed to histidine. In a similar way, the second codon, TCT, gets transcribed into the amino acid serine. And so on. The resulting protein is a chain containing all those amino acids—so hemoglobin is a chain containing histidine, serine, and so on.

Okay, so far, so good: DNA can be used as a recipe for generating proteins. Proteins, however, differ from DNA in that they each have their own special shape, unlike the completely regular structure of DNA. That shape is tremendously important. For example, as I mentioned before, the antibodies in our immune system are proteins, and the shape of an antibody determines which viruses it can lock onto. What's going on is that as the information in the DNA is transcribed to form the amino acids in the protein, the protein "folds" into its shape. How this folding occurs is still only partially understood, but there are some basic rules of thumb that should give you the flavor of what's going on. Some amino acids like to be near water—they're called *hydrophilic*, from the Greek roots "hydro" and "philia," for water and love, respectively. Since proteins inside a cell are surrounded by water, the protein will tend to fold so the hydrophilic amino acids sit on the outside, near the water. Histidine and serine are both examples of hydrophilic amino acids. By contrast, *hydrophobic* amino acids—amino acids that don't like water—end up bundled up tight inside the protein. Sometimes these tendencies conflict: neighboring amino acids in the protein may be alternately hydrophobic and hydrophilic, with the result that the protein can end up folding into a very complex shape.

There's an incredibly clever trick here that nature is using. The DNA is a completely regular arrangement of information, which makes it both easy to copy and relatively straightforward to transcribe into amino acids. But then competition between hydrophilia, hydrophobia, and other forces means that the protein can fold up to form complex shapes. By changing the DNA we can change the amino acids in the protein, which in turn causes the shape of the protein to change. What's clever about this is that it takes us from the regularly arranged information in the DNA, which is easily copied, to the many possible shapes of the protein. A priori, shapes don't seem so easy to copy. It's as though you could trace over the blueprint for a house, and the traced version would then somehow spring into existence as a tiny model house. The DNA-protein connection is Nature's way of making easy the seemingly impossible task of copying complex shapes.

But there's a problem with this neat story. Just because we know the DNA sequence for a protein doesn't mean we can easily predict what shape the protein has, or what the protein will do. In fact, today we have only a very incomplete understanding of how proteins fold. Complete structures—the exact shapes—are known for only 60,000 proteins, despite the fact that we know the DNA sequences for millions of proteins. Most of those complete structures have been found using a technique called *X-ray diffraction*—basically, shining X-rays at a protein and figuring out its shape by looking carefully at the X-ray shadow it casts. It's slow, expensive, painstaking work, and the techniques are only gradually getting better. What we'd really like is a fast and reliable way to predict the shape from the genetic description. If we could do that, cutting out the slow and expensive X-ray diffraction step, we'd go from knowing the shape of 60,000 proteins to knowing the shape of millions. Even more significantly, such a method would be a tremendously powerful tool for helping us design proteins with desired shapes. This would, for instance, help us engineer new antibodies to fight disease.

To solve the protein folding problem, biochemists have turned to computers in an attempt to predict protein shape from the genetic description. To make their predictions they use the idea that a protein will eventually fold into its lowest energy shape, much as a ball will roll to the bottom of a valley between two hills. All that's needed is good method for finding the lowest energy shape of a protein. This sounds promising, but in practice it's hard to search through all the possible shapes, looking for the shape with the lowest energy. The difficulty is the number of different shapes a protein can potentially fold into. Proteins typically have hundreds or even thousands of amino acids. To determine the structure means knowing the exact position and orientation of every single one of those amino acids. With so many amino acids involved, the number of possible shapes is astronomical, far too many to search through even on a very powerful computer. Enormous effort has been put into finding clever algorithms that can be used to restrict the number of configurations that must be examined, and the algorithms are getting pretty good. But there's still a long way to go before we can use computers to reliably predict protein shapes.

In 2007, a biochemist named David Baker and a computer graphics researcher named Zoran Popović, both from the University of Washington, in Seattle, had an idea for a better way of solving the problem. Baker and Popovió's idea was to create a computer game that shows a protein to the player, and gives them controls to change the shape, rotating the protein, moving amino acids around, and so on. Some of the controls built into the game are similar to the tools used by professional biochemists. The lower the energy of the shape the player comes up with, the higher their score, and so the highest scoring shapes are good candidates for the real shape of the protein. Baker and Popović hoped that this might be a better approach to protein folding than the conventional approaches, combining state-of-the-art computational techniques with computer gamers' persistence and abilities at pattern matching and 3-D problem solving.

I was skeptical when I first heard about Foldit. It sounded like the dull educational computer games I saw in school when I was growing up in the 1980s. But I downloaded the game, and spent hours playing it over several days. At that point, the excuse "I'm doing research for my book" was rapidly becoming a euphemism for "this is a great way to procrastinate on writing my book," and I forced myself to stop. So far, more than 75,000 people have signed up. People play the game because it's good. It has the compelling, addictive quality all good computer games have: a task that's challenging but not impossible, instant feedback on how well you're doing, and the sense that you're always just one step away from improvement. It's the same addictive quality we saw earlier in the MathWorks competition, and which is also felt by many participants in Galaxy Zoo. Furthermore, like Galaxy Zoo, Foldit is deeply meaningful to many of the players. Einstein once explained why he was more interested in science than politics by saying, "Equations are more important to me, because politics is for the present, but an equation is something for eternity." Each time you classify a galaxy or find a better way to fold a protein, you're making a small but real contribution to human knowledge. For many participants, Foldit and Galaxy Zoo aren't guilty pleasures, like playing World of Warcraft or other online games. Instead, they're a way of contributing something important to society. One of the top Foldit players, Aotearoa, describes it as "the most challenging, exciting, stimulating, intense, addictive game I have ever played," and comments that it provides a way for people to "offer something proactive to solving some of the worlds/societies most complicated puzzles, rather than waste time playing a 'game' that does not provide the same 'rewards' as folding protein does, this way!"

In addition to the individual motivation to play, Foldit also encourages collective problem solving by the players. There is an online discussion forum and a wiki, where players share news and discuss their strategies for protein folding. The game incorporates a simple programming language that players can use to create scripts—short programs—that automate game tasks. A typical script might implement a strategy for improving a fold, or identify which part of the protein's current shape is in most need of improvement. Hundreds of such scripts have been publicly shared—an open source approach to protein folding. Many of the players work in groups, sharing their insights about the best ways of folding. All this work is greatly informed by the game score, which, as in the MathWorks competition, focuses participants' attention where it will be most useful: when one of the high-scoring players shares a strategy tip or a script, other players pay attention. The players themselves are wildly varied, ranging from a self-described "educated redneck" from Dallas, Texas, to a theater historian from South Dakota, to a grandmother of three with a high-school education.

Just how good are the Foldit players at folding proteins? Every two years since 1994, there's been a worldwide competition of biochemists using computers to predict protein structures. The competition, called CASP—Critical Assessment of Techniques for Protein Structure Prediction—is very important to the scientists who work on protein structure prediction. Before the competition starts, the CASP organizers approach some of the facilities that determine protein structure using the traditional approach of X-ray diffraction, and ask them what protein structures they expect to complete in the next couple of months. They then use those proteins as puzzles in

CASP. Starting with the sequence of amino acids making up the protein, the CASP competitors are asked to predict the structure. At the end of the competition, teams are ranked on how close they come to the actual structure.

Foldit players competed in both the CASP 2008 and 2010 competitions. They performed extremely well, finishing near or at the top on many of the CASP challenges. Foldit developer Zoran Popović summed up the results of the 2008 competition by saying that "foldit players are on a par, but not better than protein folding experts at trying to solve the same problem with all tools available to them. It also appears that foldit outperformed all fully automated server submissions." Thus, a team of amateurs can be competitive with some of the world's top biochemists, equipped with state-of-the-art computers. Popović told me that his "ultimate goal is to show that experts are unequivocally inferior to the general population with this problem … a biochemistry PhD does not self-select for spatial reasoning. Structure prediction is all about 3d problem solving and very little about biochemistry." Indeed, even specialists in protein-structure prediction usually spend only a small fraction of their time working directly on predicting protein structures. And while they have expertise that the amateurs don't, much of that knowledge is incarnate in the mechanics of the game. That levels the playing field enough that the remaining disparity in expertise can be overcome by the greater time commitment of the Foldit players. It's a symbiosis: the professionals develop the systematic understanding that underlies the mechanics of the game, and the amateurs then supply the dedicated artistry required to take best advantage of that systematic understanding.

Citizen Science Today

Citizen science is not an invention of the internet era. Many of the earliest scientists were amateurs, often pursuing science as a hobby alongside some more lucrative profession, such as astrology. But even after science was professionalized, amateurs continued to dominate some parts of science. For example, many of history's most successful comet hunters have been amateur astronomers, people such as John Caister Bennett, a civil servant in the South African city of Pretoria, who discovered one of the most spectacular comets of the twentieth century, the great comet of 1968, Comet Bennett.

Although citizen science is not new, online tools are enabling far more people to participate—think of Galaxy Zoo's 200,000-plus participants and Foldit's 75,000-plus participants—and also expanding the range of scientific work those people can do. To be a comet hunter in the 1960s you needed to purchase or build a telescope, learn how to use it, and then spend many, many hours observing the sky. The barriers to entry and to continued contribution were high. By contrast, you can get started on Galaxy Zoo or Foldit in a matter of minutes. It's even possible to classify galaxies on your smartphone. Aside from dropping barriers to entry, online tools also enable sophisticated interactive training, and bring participants together in communities where they can learn from one another, and support one another's work. As a result we're seeing a great flowering of citizen science.

As an example of this flowering, comet hunting has been transformed by the internet. In 1995 the European Space Agency and NASA launched a spacecraft called SOHO, which was designed to take exceptionally good photos of the sun and its immediate neighborhood. (SOHO stands for Solar and Heliospheric Observatory.) It turns out that near the sun is a great place to look for comets, in part because comets are very well illuminated there, and in part because their tails are elongated by the solar wind. Ordinarily such comets wouldn't show up in photos because of glare from the sun, but one of the instruments on SOHO is specially designed to block out light from the sun's main body so that it can take photos of the sun's corona—the plasma "atmosphere" just above the sun's surface. The SOHO team decided to share their images of the corona openly on the internet, and many amateur comet hunters began combing through the photos, looking for comets. The most successful is a German amateur astronomer named Rainer Kracht, who spends hours each week looking very, very carefully at pictures from SOHO. In this way he has become the most successful comet hunter in history, so far discovering more than 250 comets, almost one in 15 of all the comets ever discovered.

Another example of citizen science is Project eBird, run by Cornell University's Laboratory of Ornithology. eBird asks amateur birdwatchers to upload information about the birds they see to an online website: what species of bird they saw, when they saw it, and where they saw it. By combining all the submitted observations, eBird can build up an understanding of the world's bird populations. This is another case where citizen science is building on an earlier tradition, this time a tradition of collaboration between amateur birdwatchers and professional ornithologists. But Project eBird is enabling this collaboration on an unprecedented scale, with participants so far reporting more than 30 million bird observations. About 2,500 birdwatchers are frequent contributors to the site, making 50 or more contributions, and tens of thousands of people regularly use the site. The data collected can be used, for example, to generate range maps showing the density of some particular species of bird in different locations. As eBird gathers more data (it began in 2002) such range maps will become increasingly useful for tracking the impact on birds of effects such as climate change, changes in nearby human population, and other environmental factors.

Yet another example of citizen science comes from the study of dinosaurs. Most dinosaur research concentrates on just one or a few fossils. In September of 2009, paleontologists Andy Farke, Mathew Wedel, and Mike Taylor had the idea of creating a large database containing information about many dinosaurs, by combining the results of hundreds or even thousands of scientific papers. Their hope was that the database could then be mined to answer many new questions. But instead of building the database on their own, they decided to harness the distributed knowledge and effort of a broader community of people. They started the Open Dinosaur Project, recruiting people from all over the world to, er, dig up papers about dinosaurs. As I write, they're focusing on dinosaur limb measurements. If a volunteers finds a paper studying, say, a *Stegosaurus* specimen with a right femur that's 1,242 millimeters in length, they would record that piece of data in the database. The project has thus created a list of measurements from 1,659 separate dinosaur specimens, contributed by 46 people, many amateurs. Their hope is that this will let them answer questions about (for example) the evolution of dinosaur locomotion. It's still early days in the Open Dinosaur Project, and while data are being collected quickly, it's too soon to say how useful the data will be. But it's another example of how a community containing both amateur and professional scientists can do more than either group could on their own.

From these and earlier examples, we see several distinct ways that citizen scientists are contributing to science. Citizen science can be a powerful way both to collect and also to analyze enormous data sets. In those data sets, citizen scientists can scout out the unusual and the unexpected, discoveries such as the voorwerp and the green peas, discoveries that would be difficult to program a computer to spot. Citizen science thus complements the tools of data-driven intelligence described in the last chapter.

Citizen scientists can also work to symbiotically extend the capability of those tools, as demonstrated by the Foldit players' artistry in using the tools of protein-structure prediction. In another twist on this idea, the Zookeepers have recently used the Zooites' galaxy classifications to train a computer algorithm to distinguish between spiral and elliptical galaxies. The preliminary results are promising, with the algorithm achieving 90 percent agreement with the human classifications. This result is interesting in part because future sky surveys from instruments such as the Large Synoptic Survey Telescope (the LSST, described on page 107) will produce vastly more data than even the huge crowd of volunteers at Galaxy Zoo can hope to analyze. Perhaps the results from the LSST will be understood by first asking amateurs to analyze a small portion of the data, and then using computer algorithms to learn from the amateurs' analyses, with computers completing the classification of the entire data set. Possibilities such as these are creating a massive efflorescence of citizen science projects, with ordinary people participating in scientific research in ways unimaginable a generation ago.

How Much Will Citizen Science Change Science?

Examples such as Galaxy Zoo, Foldit, and the open dinosaur project are interesting and fun. But science is vast, and while citizen science is likely to grow rapidly in the years and decades ahead, that does not mean that it will come to be a dominant part of how science is done. Although projects such as Galaxy Zoo are important, it's not obvious whether they're curiosities or harbingers of a broader change in science. Will citizen science ever have a broad and decisive impact on how science is done? Or is it destined to be useful mainly in a few particular corners of science? I don't know the answer to these questions. We've only just begun exploring the ways online tools can expand the impact of citizen science. The situation is quite different from the changes described in the last chapter. There, as we saw, powerful new tools for finding meaning in knowledge are already revolutionizing many parts of science. As yet, the prospects for citizen science are more uncertain. But although we can't know for sure how important citizen science will ultimately be, we can at least think a little more about its potential, where it might be applied, and what its limitations might be.

Part of that potential is to create supportive and stimulating communities for citizen science. Before the internet, most citizen scientists worked largely on their own, isolated from the encouragement and criticism of colleagues. Today, that's changing. In the Galaxy Zoo forums you see a community where people help out one another, a supportive environment in which they can learn and grow as astronomers, a place where people can ask questions and other people will answer in a friendly way. Consider, for example, the way the Zooites helped each other in their quest to understand the green pea galaxies. They repeatedly critiqued and improved one another's ideas about what made the green peas unique, egging one another on, sharing tidbits about problems such as how best to analyze a galaxy's spectrum, or how to do database queries to automatically find green peas in the SDSS data. When you're in a community like that, you're getting constant feedback that says, in effect, "Hey, this is important, this is what really *matters*." Think of the way children play soccer or baseball in streets and parks—they play tirelessly, hour after hour, day after day, gradually getting better as part of a community that both demands their best, and makes reaching it a joy. All the most creative communities do the same.

This new type of community building is important, but today's citizen science projects have a great deal of room for improvement. Galaxy Zoo, Foldit, and most other citizen science projects don't yet have the kind of structured stepping stones of development and mentorship available to professional scientists, stepping stones that help those scientists acquire the broad base of knowledge required for many types of scientific work. It will be interesting to see how citizen science projects evolve. Will we see ever more effective learning environments, a place where amateurs can learn as they go, gradually acquiring more expertise? Will we see systems of mentorship emerge, giving people a structured way of learning? Imagine online communities built around virtual seminar series and conferences, online question-and-answer sessions, and discussion groups. These and other ideas can be used to create a demanding and rewarding online community supporting citizen science.

The biggest citizen science projects have recruited large numbers of people—Galaxy Zoo has more than 200,000 participants—and you might wonder if there is much more room for citizen science to grow. Or has the public appetite for citizen science already been exhausted? There's a nice way of thinking about these questions, inspired by an analysis of the analogous questions for Wikipedia done by the author Clay Shirky, of New York University. To start, let's figure out a rough estimate of the total effort involved in a project such as Galaxy Zoo. So far, the Zooites have done approximately 150 million galaxy classifications. If each classification takes, say, 12 seconds, then that works out to 500 thousand hours of work. That's like having 250 employees work full time for a year! While this is an amazing amount of work, on the scale of society as a whole it's a drop in the bucket. On average Americans watch five hours of television per day, which over the course of a year means that Americans are watching more than 500 billion hours of television. That's a million Galaxy Zoo projects!

Let's look at an activity that's closer to Galaxy Zoo in scale. The English soccer club Manchester United seats 76,000 at their home stadium, Old Trafford. Games take two hours, with stoppages, so the spectators at a game are spending roughly 150,000 hours of time in total, nearly a third of the amount of time the Zooites have spent classifying galaxies! To put it another way, imagine that you filled up the Manchester United stadium, and instead of watching soccer, you asked people to classify galaxies for a couple of hours. If you did that three times, then you'd roughly match the effort put into Galaxy Zoo. Of course, Galaxy Zoo has been running three years as I write, while Manchester United plays dozens of home games each year. So the Zooites are a notch or two down from the devotion shown by Manchester United's home game fans. A closer comparison is to a much smaller soccer club, such as the Bristol Rovers, who get a few thousand fans to each home game. There's a great deal of room for citizen science to grow!

Shirky has coined the phrase "cognitive surplus" to describe our society's disposable time and energy—all the time we collectively have when we're not dealing with the basic obligations of life, such as making a living or feeding our family. It's the time we put into leisure activities such as watching television, or going out with friends, or relaxing with a hobby. Mostly, these are activities we do individually or in small groups. What the online tools do is make it easy to coordinate complex creative projects in a large group. It's always been possible for a large group of people to get together and cheer at a soccer game. But it's much harder to get a large group of people together to work toward a complex creative goal. One way is to pay all those people to come together and form a hierarchy organized into managers and subordinates. We call that a company or a nonprofit or a government. But without money it's historically been difficult to hold such complex creative projects together. Online tools make it much easier to do this complex coordination, even without money as a motivator. As Shirky poetically puts it:

> We are used to a world where little things happen for love and big things happen for money. Love motivates people to bake a cake and money motivates people to make an encyclopedia. Now, though, we can do big things for love.

Projects such as Galaxy Zoo and Foldit are doing just that, using our society's cognitive surplus to solve scientific problems.

How much of our society's cognitive surplus will ever be used to do citizen science? Today it's not possible to answer that question. Citizen science is in the early days of a major expansion enabled by online tools. How far it ultimately expands will depend upon the imagination of scientists in coming up with clever new ways to connect with laypeople, ways that inspire them and help them make contributions they find meaningful. You get a glimpse of this in the story of one of the most prolific participants in Galaxy Zoo, a woman named Aida Berges. Berges is a 53-year-old stay-at-home mother of two originally from the Dominican Republic, now living in Puerto Rico. She classifies hundreds of galaxies every week, more than 40,000 galaxies in total thus far. She's worked on the hunt for green peas, for voorwerps, for merging galaxies, and many other projects. She's discovered two hypervelocity stars, stars which are moving so fast that they are actually leaving our galaxy. Fewer than twenty such stars have been discovered, ever, in total. Ms. Berges joined Galaxy Zoo after reading about it online and said of the experience that "my life changed forever ... it was like coming home for me."

Cynics will say that most people aren't smart or interested enough to make a contribution to science. I believe that projects such as Galaxy Zoo and Foldit show those cynics are wrong. Most people are plenty smart enough to make a contribution to science, and many of them are interested. All that's lacking are tools that help connect them to the scientific community in ways that let them make that contribution. Today, we can build those tools.

Changing the Role of Science in Society

After Jonas Salk announced his polio vaccine in 1955 it was quickly pressed into widespread use in the rich developed countries, and polio rates plummeted. But in developing countries it was a different story. In 1988, roughly 350,000 people in the developing world became infected with polio. In that year the World Health Organization (WHO) decided to launch a global initiative to wipe out the disease. They made quick progress, and in 2003 there were only 784 new cases worldwide, most concentrated in just a few countries. Worst hit was Nigeria, where nearly half (355) of the new cases occurred. WHO decided to launch a major vaccination program in Nigeria, but the initiative was blocked by political and religious leaders in three northern Nigerian states—Kano, Zamfara, and Kaduna—with a total population of 18 million people. Leaders in those states warned that the vaccines could be contaminated with agents causing HIV/AIDS and infertility, and told parents they should not allow their children to be vaccinated. The government of Kano described their opposition to vaccinations as "a lesser of two evils, to sacrifice two, three, four, five even ten children to polio [rather] than allow hundreds of thousands or possibly millions of girl-children likely to be rendered infertile." The leader of the powerful Kano State Sharia Supreme Council said the polio vaccines were "corrupted and tainted by evildoers from America and their Western allies." Vaccinations were suspended in Kano, and a new polio outbreak occurred, spreading to eight neighboring countries, and eventually causing 1,500 children to become paralyzed.

Polio vaccination is far from the only issue where good science doesn't necessarily lead to good public health outcomes. In the United Kingdom, use of the measles-mumps-rubella vaccine dropped sharply in the early 2000s after a 1998 paper in the prestigious medical journal *The Lancet* suggested the vaccine might cause autism in children. (The paper's methodology was flawed, and it was later retracted by the journal and most of the authors). The supposed vaccine-autism link became a topic of great public controversy in the UK, with Prime Minister Tony Blair publicly supporting the vaccine, but refusing to confirm whether his son Leo had been vaccinated. The vaccination rate dropped from 92 percent to 80 percent. That may sound like a small drop, but the number of measles cases jumped dramatically, rising seventeen-fold over just a few years. To understand why the increase in measles was so dramatic—and therefore why a drop in vaccination rates is such a big deal—notice that the fraction of people *not* being vaccinated rose from 8 percent to 20 percent. Roughly speaking, that meant that someone infected with measles would be exposed to two and a half times as many susceptible people as before. And if any of those people caught measles, they would, in turn, be exposed to two and a half times as many susceptible people as before. And so on. That's why even a small drop in vaccination rates can cause a big increase in disease incidence.

Vaccine fiascoes notwithstanding, our society often does a good job converting science into social good. Markets and entrepreneurship, for example, are powerful institutions that often help turn science into goods that enhance our lives. Think of a development such as lasers. When lasers were first invented, many people regarded them as toys with few apparent uses. But entrepreneurs figured out ingenious ways of using lasers to do everything from playing movies (DVDs) to correcting vision by laser eye surgery. As a society we're very, very good at taking science and using it to develop new products for delivery to market.

But while we're good at delivering science to market, we have a more mixed record when it comes to delivering science through public policy. In a market, everyone can decide for themselves whether they want to use a product. If laser eye surgery makes you squeamish, no one's making you get it. But policy decisions are often collective decisions, like whether child vaccination should be mandatory. Such decisions can't be made individually, as in a market, but require broad agreement to be effective. And when scientists discover something with dramatic policy implications— say, that human carbon dioxide emissions are leading to a warming of the global climate—then in many ways they're treated as just another interest group trying to lobby the government. But science isn't just an interest group. It's a way of understanding the world. Ideally, our institutions for governance would incorporate in public policy the knowledge gained by science—as imperfect, uncertain, and provisional as that knowledge is—as well as possible. But in today's democracies, that's not what happens. This is the problem of science in democracy.

I don't have solutions to the vaccine problem or, more broadly, to the problem of science in democracy. I'm describing these problems because they're concrete examples of critical flaws in the role science currently plays in our society. Any fix to these and similar problems will require big changes in the role of science in society. Most of the time such changes occur only very slowly, and so it's tempting to take that role for granted, to view it as a natural state of affairs. But, in fact, the current state of affairs is not natural at all: the role of science has been radically different in different societies and at different times—just think, for example, of all the societies in which scientific thought has been entirely suppressed. Historically, big changes in the role of science have often been driven by new technologies and the new institutions they enable. Think of the printing press's role as an enabler of the Renaissance, the Reformation, and the

Enlightenment. We can change the role of science in society if we change the institutional answers we give to fundamental questions such as "Who funds science?" or "How is science incorporated into government policy?" or even "Who can be a scientist?"

As a concrete example of the way institutions impact the role of science, let's return to the market system. The importance of the market to the role of science is vividly illustrated by what happened when the market was suppressed in the Soviet Union. Although the Soviet Union had one of the best scientific research systems in the world, without a market system it was mostly unable to make scientific innovations available to its citizens. Another example of the power of institutions is the way the introduction of compulsory schooling has increased general scientific literacy. Although it's conventional wisdom in many circles to complain about standards of scientific literacy, by historical standards we live in an enlightened age. Both the market and schools act as bridging institutions, connecting science to society in a way that brings many social benefits. As a final example, this time a negative example, consider the suppression of science by the early Christian Church. This lasted more than a millennium, from the Christian emperor Justinian's closing of the Academy in Athens in 529 CE to the trial and house arrest of Galileo in 1633 CE.

By changing our society's institutions, we can dramatically change the role of science in society, and perhaps address some of our society's most significant problems. To do this will require the imagination and will to invent new institutional mechanisms that could address problems such as the vaccine problem or the problem of science in democracy. It may seem unrealistic to change our institutions in this way. Most of the time institutions change only very slowly. But, today isn't most of the time. Online tools are institution-generating machines. Examples such as Galaxy Zoo, Wikipedia, and Linux demonstrate how much easier it has become to create new institutions, and even to create radically new types of institution. At the same time, online tools are transforming existing institutions in our society—consider the collapse of traditional music and newspaper companies over the past ten years, and the gradual rise of new models in their place. And so we're at a very interesting point in history, one where it's become far easier to create new institutions and to reinvent existing institutions. This doesn't mean that we can easily solve problems such as the vaccine problem. What it does mean is that we have an opportunity to reimagine and to some extent recreate the role of science in society. We're already beginning to see this happening, with citizen science projects such as Galaxy Zoo and Foldit showing how online tools can be used to change something very fundamental: who can be a scientist. In the remainder of this chapter, we'll explore other ways online tools change the role of science in society, including ways they improve public access both to the results of science, and to scientists themselves.

Open Access

Imagine you're a woman who has gone to the doctor for a regular mammogram screening, and your doctor has come back with surprising and terrible news: you have early-stage breast cancer. Shocked, you go home, and begin planning your attack on the disease. You decide that the first thing to do is to become better informed. You read around online, and discover a great deal of useful information from sites such as the cancer.gov site run by the US National Cancer Institute. But after a while, all the introductory information you find on the web becomes repetitive. You want more up-to-date knowledge on the most promising current research. A friend mentions that Google has a special search engine—called Google Scholar—which will help you search the scientific literature for the best and latest papers on breast cancer. You go to the site, search on "breast cancer," and discover umpteen-thousands of papers. Excellent! Even better, Google Scholar orders the results according to Google's best guess as to their importance. You go to download the paper Google ranks as the top result, and discover that you need to pay 50 dollars for the download. "Never mind," you think, "I'll come back to that paper later." But when you look at the second paper, you discover it costs 15 dollars to download. Onto the third paper, and that publisher wants to charge you, too, but is coy about the price, asking you to register on their site first. As you continue paging through the results, the pattern of fees continues, and your initial elation turns to angry disbelief. "Surely," you reason, "with tens of billions of dollars of taxpayer money spent each year on scientific research, we should at least be able to read the results of that research?!" Now, breast cancer is a serious disease and you're tempted to swallow your anger and pay the fees. But there are thousands of papers. There's no way you can afford to pay for even a tiny fraction of them.

Traditional scientific publishing is based on a pay-for-access model. In many ways it works much like the magazine business, and there's less difference than you might think between a leading science journal such as *Physical Review Letters* and magazines such as *Time* and *People*. Like the magazines, science journals are collections of articles, but instead of discussing news, politics, and celebrities, the journal articles describe scientific discoveries. Journals may not have flashy covers and advertising, nor will you find most of them on display at your local newsstand, but both journals and magazines make much of their money by charging readers. An annual journal subscription might run to hundreds or thousands or even tens of thousands of dollars. And, as we've just seen, journals supplement those fees by charging for one-off access to articles on the web, typically $10 to $50.

This subscription-based business model has been used by scientific publishers for hundreds of years. It's a model that has served both science and society well. But the internet makes it possible to move to a new model of *open access* to scientific papers, where those papers may be freely downloaded. This is part of the shift we saw in the last chapter, with all the world's scientific knowledge gradually becoming accessible online. A caveat to that story, though, is that at present much of the knowledge is only accessible *if you're a scientist*. In particular, scientists often work at universities that have bulk subscriptions to thousands of scientific journals. A scientist can freely download as many articles about breast cancer or any other subject as they wish, while other people are kept out by the fees. It's as though there is a wall dividing humanity. On one side of the wall are 99-plus percent of the human beings who have ever lived. And on the other side of the wall is the world's scientific knowledge. The *open access movement* is trying to break down that wall. Just as citizen science is changing who can be a scientist, the open access movement is changing who has access to the results of science.

One of the standout successes of the open access movement is a popular website known as the physics preprint arXiv (pronounced "archive"). A "preprint" is a scientific paper, often at late draft stage, ready to be considered by a scientific journal for publication, but not yet published in a journal. You can go to the arXiv right now, and you'll find hundreds of thousands of up-to-the-minute preprints from the world's physicists, all available for free download. Want to know what Stephen Hawking is thinking about these days? Go to the arXiv, search on "Hawking," and you can read his latest paper— not something he wrote a few years or decades back, but the paper he finished yesterday or last week or last month. Want to know the latest on the hunt for fundamental particles of nature at the Large Hadron Collider (LHC)? Go to arXiv, search on "LHC," and you'll get a pile of papers to make you stagger. If you get a kick out of surprising people, it might make for unusual cocktail party conversation: "So, did you see the latest on the LHC's hunt for the Higgs particle? Turns out..." Of course, it's not all easy reading. Many of the papers are written by physicists for physicists, and they can get extremely technical. But even the most technical papers often have intriguing nuggets that are accessible to the layperson.

The arXiv site works like this. When a physicist completes their latest paper, they go to the arXiv website and upload it. A quick check is done by arXiv moderators to remove inappropriate submissions— you won't see Viagra advertisements or too many obviously crackpot papers. A few hours later the paper appears on the site, where it can be downloaded and read by anyone in the world. Many physicists submit their papers to the arXiv as soon as they're complete, and long before they're published in a conventional scientific journal. More than half of all papers in physics appear in the arXiv, and in some subfields of physics the fraction is nearly 100 percent. Many physicists begin their working day by checking the arXiv to see what appeared overnight. It's revolutionized physics, by speeding up the rate at which scientific discoveries can be shared. At the same time, the arXiv has made much of humanity's knowledge about physics freely accessible to anyone with an internet connection. Whether or not you have any personal interest in physics, it's to society's great benefit that this knowledge is freely available to entrepreneurs and engineers, to journalists and students, and to many others who can benefit, but who were formerly locked out.

The arXiv is one of the big successes of the open access movement. But in most fields of science, fields such as medicine, climate science, and the environment, humanity's scientific knowledge is still mostly accessible only to scientists, and to whoever else can pay for access. Because of this, and inspired in part by the success of the arXiv, several organizations are creating open access solutions for fields other than physics. An example is the Public Library of Science, or PLoS. Founded in 2000, PLoS is in many respects more like a traditional journal publisher than it is like the arXiv. But rather than charging readers for access to papers, PLoS instead charges authors to publish their papers. That charge funds PLoS's operation, making it possible for PLoS papers to be made freely available on the web. Using this model, PLoS has rapidly built journals regarded as among the best in their fields, journals such as PLoS Biology and PLoS Medicine.

The arXiv and PLoS are just two of many efforts aiming to make open access to the scientific literature the norm. Many other open access projects have been launched. These projects have been gaining traction, and in 2008 the US Congress signed into law the National Institutes of Health (NIH) Public Access Policy. The NIH policy requires anyone funded by the NIH to upload finished papers to an openly accessible archive within 12 months of publication in a conventional journal. With a budget of more than 30 billion dollars per year, the NIH is the world's largest scientific grant agency, and so this policy is rapidly increasing the amount of openly accessible research. Many other grant agencies and universities around the world are implementing similar open access policies. For instance, all of the UK Research Councils now have policies along similar lines to NIH's requiring researchers to make their papers openly available. Although much scientific research still remains locked behind publisher paywalls, we may be on the verge of a major shift toward open access as the norm, not the exception. If that happens, people in the decades to come will look back in amazement that there was ever a time we did not have universal access to science. It will be an institutional shift not unlike the introduction of the market.

The most obvious benefit of widespread open access is to individual citizens: no more restrictions on the ability of people suffering diseases to download the latest research! But over the long run an even bigger benefit of open access will be that it enables the creation of other institutions bridging science and the rest of society. We're already starting to see this happen. For example, user-generated online news sites such as Digg and Slashdot routinely link to the latest research in the arXiv and PLoS and other open access sources. These news sites enable ordinary people to collectively decide what the news is, and provide a space where they can discuss that news. Often, what people choose to discuss includes the latest papers at the arXiv on subjects such as cosmology and quantum teleportation, or the latest papers at PLoS on subjects such as genetics and evolutionary biology. When people on the news sites post links to pay-for-access journals such as *Nature* and *Science*, complaints often ring out, and users sometimes point out pirated online copies as an alternative. (This is not something I endorse, but it does happen!) In a similar way, professionally produced online news sites such as ScienceNews offer their perspective on the latest research. They cover both open and closed access stories, but the open access stories often get more attention simply because people can click through to see the original research. These sites provide a window on the scientific community, complementing and extending resources such as the arXiv and PLoS. Of course, the effect of these changes is at times mixed. Many news articles have been written about papers of dubious scientific merit that have appeared on the arXiv and other open access resources. But insofar as scientists have good faith evidence on their side, open access is a powerful platform for building new institutions for the betterment of society.

The reactions of traditional pay-for-access scientific publishers to open access have varied. Some have begun their own experiments with open access. But many, including some of the largest publishers, feel threatened by the open access movement. For them, open access archives and journals aren't run-of-the-mill business competitors. Instead, they have the potential to radically change the business model of scientific publishing. Traditional publishers face a tough choice. Should they adopt the open access model of PLoS and journals like it? Or should they stay as they are? Should they go even further, and fight against open access, for instance by lobbying against policies such as the NIH open access policy? It's a difficult choice to make, for if they go the open access route, it's possible that it will greatly reduce journal revenues. Unless those companies develop new sources of revenue, their employees will lose jobs, and shareholders will lose money. That's tough to face after decades and sometimes centuries of hard work building businesses that have served society well. But society's best interest has shifted away from that old business model. It's no wonder many traditional publishers feel threatened. The available technology may have changed, but that doesn't mean the business models have.

Monetarily, there's a lot at stake here: scientific publishing is a big business. This may be a surprise to you. Certainly, when it comes to high-flying professions, not many people think of scientific journal publishing. CEOs from scientific publishers don't often appear on the cover of *Forbes* or *Business Week*, alongside software moguls or hedge fund operators. But maybe they should, because scientific publishing is staggeringly profitable. The world's largest scientific journal publisher is the company Elsevier. In 2009 Elsevier made a profit of 1,100 million US dollars, more than a third of their total revenue of 3,200 million dollars. As a share of revenue, that's the kind of profit enjoyed by businesses such as Google, Microsoft, and a very few others. Elsevier is so profitable that its parent company, the Reed Elsevier Group, recently sold off another big part of their business, the educational publisher Harcourt, for close to five billion dollars, to help finance the expansion of Elsevier's journal publishing business. And while Elsevier is the biggest of the scientific publishers, many other scientific publishers also do amazingly well. Even some not-for-profit scientific societies make a lot of money by publishing journals for their members, with the profits then subsidizing other society activities. For example, in 2004 the American Chemical Society made a profit of about 40 million US dollars on their journals and online databases, out of revenue of 340 million dollars. That's much less than Elsevier, but remember: this is a not-for-profit society!

With so much at stake, it's no surprise that some traditional scientific journal publishers have begun aggressively lobbying against open access. According to a report published by *Nature* in 2007, a major publishers' trade association hired high-priced public relations consultant Eric Dezenhall to help them take on the open access movement. Dezenhall had earned a reputation as the "pit bull" of the public relations world, with clients including Jeffrey Skilling, the disgraced former Enron chief, and ExxonMobil, which hired Dezenhall's company to help them take on Greenpeace. Dezenhall advised the publishers to focus on simple messages such as "Public access equals government censorship," and suggested that they try to "paint a picture of what the world would look like without peer-reviewed articles." (Both notions are false: open access doesn't involve censorship, nor does it mean giving up peer review.) When asked about the move to hire Dezenhall, a vice president at the publishers' association replied, "It's common to hire a PR firm when you're under siege." Not long after receiving Dezenhall's advice, the publishers' association launched an organization called PRISM, the Partnership for Research Integrity in Science and Medicine. PRISM began a publicity initiative arguing against open access policies such as the NIH policy, claiming that open access would threaten "the economic viability of journals and the independent system of peer review" and potentially introduce "selective bias into the scientific record."

The Dezenhall-PRISM story is just one skirmish of many in the battle between some traditional scientific publishers and the open access movement. On the one hand, we have a situation where open access poses a threat to the profits and ultimately the jobs of both the traditional scientific publishing companies and the not-for-profit scientific societies. But balanced against this is a marvelous opportunity: as examples such as the arXiv and PLoS and the NIH open access policy show, it's now feasible to make all scientific knowledge freely available to all of humanity. And that will bring astonishing benefits, benefits far too great to refuse merely to preserve a few successful businesses. As occurs so often with the introduction of new technologies, we are weighing a great good for society against harm for a few. The traditional publishers who are battling against open access should have our sympathy, but not our support.

Science Blogging

In April of 2008, author Simon Singh wrote a piece in the *Guardian* newspaper, where he criticized the British Chiropractic Association

(BCA) for claiming "that their members can help treat children with colic, sleeping and feeding problems, frequent ear infections, asthma and prolonged crying, even though there is not a jot of evidence. This organization is the respectable face of the chiropractic profession and yet it happily promotes bogus treatments." The BCA responded by suing Singh under UK libel laws, claiming that the effectiveness of its treatments was supported by a "plethora of evidence." The case received a lot of public attention in the UK, and fourteen months after Singh's article the BCA released a seven-page document describing evidence for the effectiveness of chiropractic treatments.

What happened next was unexpected. Almost immediately, the evidence released by the BCA was investigated and torn apart by an ad hoc group of science bloggers, acting on their own initiative. Here's how the events were described in an article in *The Lawyer* written by Robert Dougans, a lawyer who acted for Singh in the case, and David Allen Green, a blogger who had been covering the case:

> In less than a day, the credibility of this evidence—and indeed that of the BCA for commending it—was destroyed. A dozen or so scientist-bloggers, including a Fellow of the Royal Society, were able to track down and assess each of the scientific papers cited by the BCA and were able to show beyond doubt that these papers did not support the BCA position at all. This was a stunning and devastating blogging exercise, and when it was formally repeated by the British Medical Journal a few weeks later it was almost an afterthought. The technical evidence of a claimant in a controversial case had simply been demolished—and seen to be demolished—but not by the conventional means of contrary expert evidence and expensive forensic cross-examination, but by specialist bloggers. And there is no reason why such specialist bloggers would not do the same in a similar case.

Dougans and Green called the process "wiki litigation," and commented that its importance to the case went well beyond demolishing the BCA's evidence. They said that blogging substantially influenced coverage of the case in the mainstream media, and also "provided Singh with varied and well-reasoned views at each stage beyond that of his legal team and campaigning enthusiasts. Singh certainly took these views into account in his decision-making." It's a remarkable demonstration of how a group of bloggers can cause change in society.

Like open access and citizen science, science blogs are an institution that is changing the role of science in society. I won't talk in all that much detail about science blogging here. The reason is that ever since blogging (in all its forms, not just science blogging) began in the 1990s, there's been a lot of brouhaha about it—I've lost track of the number of magazine and newspaper articles I've seen saying "Blogs are revolutionizing politics!"; "And journalism!"; "No, they're not!"; and so on. I don't want to cover that well-trodden ground again. But I do want to describe a few examples giving the flavor of how science blogs can establish a new type of relationship between the scientific community and the broader community, complementing and extending ideas such as open access.

One remarkable aspect of the most widely read science blogs is their popularity. Pharyngula, a blog run by biologist Paul Myers from the University of Minnesota, receives over 100,000 visits per day, comparable to the circulation of a leading daily newspaper in a large metropolitan center such as the *Des Moines Register* or the *Salt Lake Tribune*. This is not bad for one guy writing in his spare time—and far more attention than all but the most famous mainstream print journalists regularly receive.

Pharyngula is the most popular science blog, but many other science blogs have thousands or tens of thousands of regular readers. My vote for the best blog in the world is the blog of Terence Tao, a Fields Medal–winning mathematician based at UCLA. (We met Tao briefly earlier, as one of the participants in the Polymath Project.) Tao's blog contains hundreds of posts. Some of the posts are lighthearted ("Quantum mechanics and Tomb Raider"), but most of the posts contain highly technical mathematics. Just to give you the flavor, posts include "Finitary consequences of the invariant subspace problem" and "The transference principle, and linear equations in primes." Although the titles look forbidding to non-mathematicians, for mathematicians these posts are remarkably clear and insightful expositions of difficult topics, often containing many thoughtful original insights. Despite its technical nature, more than 10,000 people read Tao's blog. The comments section reveals that while many of these people are professional mathematicians, many are also students, sometimes in remote locations. Some of the commenters have little mathematical background: they are just interested people who wish to learn more and who enjoy being exposed directly to the thinking of one of the world's leading scientists.

What should we make of science blogging? Is it going to transform the world? In its current form, I don't think so. Instead, the way to think about science blogging is as a harbinger of what's possible. Science blogs show in nascent form what can happen when you remove the barriers separating scientists from the rest of the community, and enable a genuine two-way flow of information. A friend of mine who was fortunate enough to attend Princeton University once told me that the best thing about attending Princeton wasn't the classes, or even the classmates he met. Rather, it was meeting some of the extraordinarily accomplished professors, and realizing that they were just people—people who sometimes got upset over trivial things, or who made silly jokes, or who made boneheaded mistakes, or who had faced great challenges in their life, and who somehow, despite their faults and challenges, very occasionally managed to do something extraordinary. "If they can do it, I can do it too" was the most important lesson my friend learned.

What's important then is that blogs make it possible for anyone with an internet connection to get an informal, rapid-fire glimpse into the minds of many of the world's scientists. You can go to the blog of Terence Tao and follow along as he struggles to extend our understanding of some of the deepest ideas of mathematics. It's not just the scientific content that matters, it's the culture that is revealed, a particular way of viewing the world. This view of the world can take many forms. On the blog of experimental physicist Chad Orzel you can read his whimsical explanations of physics to his dog, or his discussions of explosions in the laboratory. The content ranges widely, but as you read, a pattern starts to take shape: you start to understand at least a little about how an experimental physicist views the world: what he thinks is funny, what he thinks is important, what he finds irritating. You may not necessarily agree with this view of the world, or completely understand it, but it's interesting and transformative nonetheless. Exposure to this view of the world has always been possible if you live in one of the world's intellectual capitals, places such as Boston, Cambridge, and Paris. Many blog readers no doubt live in such intellectual centers. But you also routinely see comments on the blog from people who live outside the intellectual centers. I grew up in a big city (Brisbane) in Australia. Compared to most of the world's population, I had a youth of intellectual privilege. And yet the first time in my life that I heard a scientist speaking informally was when I was 16. It changed my life. Now anyone with an internet connection can go online, and get a glimpse into how scientists think and how they view the world, and perhaps even participate in the conversation. How many people's lives will that change?

Imagining New Institutions

Institutions such as citizen science, open access, and science blogging are all changing science's role in our society. Today, these institutions are small, but they're growing rapidly. Although events such as the Singh case and Hanny's discovery of the voorwerp are significant, their impact is tiny when compared to society's largest institutions, such as compulsory schooling. But most big and important institutions start out tiny and inconsequential—think of the humble origins of the school system, or of democratic government. What matters is not the absolute size of an institution, but rather its potential to grow. Institutions are what happens when people are inspired by a common idea, so inspired that they coordinate their actions in pursuit of that idea. Online tools make it far easier to create institutions, by amplifying ideas faster than ever before, and by helping coordinate action.

As an example, Galaxy Zoo began in 2007 with two guys in a pub, working on a budget of chutzpah and imagination. Three years later it involved 25 professional astronomers and 200,000 amateurs. It's expanded to include projects such as Moon Zoo and Project Solar Storm Watch. How much larger will it be in ten years' time? Suppose Galaxy Zoo decides to systematically solicit proposals from the astronomy community for the analysis of data sets. It's not too much of a leap to imagine Galaxy Zoo becoming an institution crucial to the whole field of astronomy, and perhaps to other fields as well. What other new institutions will we have the chutzpah and imagination to dream up? What other new answers will we find to fundamental questions about the role of science in society?

Bridging the Ingenuity Gap

The most isolated place in the world is Easter Island. It's a tiny island in the southeast Pacific, just 25 kilometers (15 miles) across, 3,500 kilometers (2,200 miles) west of Chile, and 2,100 kilometers (1,300 miles) east of the Pitcairn Islands. The island was originally settled by Polynesian islanders, and its culture thrived for hundreds of years, with the population growing to somewhere between 10,000 and 30,000 people. But as the population grew, the islanders consumed more and more of the island's resources, and sometime in the 1500s or 1600s, its society collapsed. When Easter Island's European discoverer, the Dutch explorer Jacob Roggeveen, arrived in 1722, he found an island stripped of natural resources. There was not a single tree higher than three meters anywhere on the island. Today, by analyzing pollen from the island, we know that Easter Island was formerly a subtropical forest, with at least 21 species of tree, some of them growing up to 30 meters high. Roggeveen also found not a single species of land bird. Today we know that at least six species of land bird used to live on the island. As the Easter Islanders destroyed their stocks of food and timber, they began to starve, and the population crashed, dropping perhaps 90 percent. Easter Island culture descended into warfare and eventually cannibalism.

The author Thomas Homer-Dixon has coined the phrase "ingenuity gap" to describe the gap in difficulty between the problems faced by a society and that society's capacity to solve problems. What happened to the Easter islanders is that they were overcome by the ingenuity gap facing their society, unable to find solutions to the problems they had created. That ingenuity gap caused the collapse of their civilization.

Modern global society faces its own ingenuity gap. We have problems such as HIV/AIDS, which reduces average life expectancy in the most highly affected African countries by 6.5 years, from 54.8 years to 48.3 years. We have the problem of nuclear weapons, with a nuclear-armed India and Pakistan arguing over Kashmir, and the world's two new superpowers, China and India, vying for supremacy in Asia. As nuclear proliferation continues, the number of plausible nuclear conflicts is rapidly rising. We face potential shortages of oil and water, and the possibility of future bio-terrorism. And, of course, there's the best-known existential threat of our time, human-caused climate change. Many of these are problems that we understand scientifically. But just because we understand the problems and their solutions at a factual level doesn't mean we can muster the collective ability to take action. We are lacking the institutional ingenuity necessary to turn our knowledge into real solutions. Today, online tools are giving us an opportunity to create new institutions to change and redefine the relationship between science and society. It is my hope that this opportunity will help us create a more resilient society and, in the memorable phrasing of Hassan Masum and Mark Tovey, bridge the ingenuity gap.

CHAPTER 8

The Challenge of Doing Science in the Open

Late in the year 1609, Galileo Galilei pointed one of his newly built telescopes up at the night sky and began to make one of the most astonishing series of discoveries in the history of science. Galileo's first major discovery, made in January of 1610, was of the four largest moons of Jupiter. Today, this discovery perhaps seems unremarkable, but it caused the biggest change to our conception of the universe since ancient times. The discovery became a sensation, and Galileo was feted throughout Europe. It also brought him the patronage of one of the wealthiest men in Europe, the Grand Duke of Tuscany, Cosimo de' Medici.

With fame and patronage came pressure to repeat his success, and Galileo wanted more discoveries to match the moons of Jupiter. He didn't have long to wait. Shortly before dawn on the morning of July 25, 1610, Galileo pointed his telescope at Saturn, and observed that it wasn't just a single round disk, as had hitherto been thought. Instead, alongside Saturn's main disk he saw two small bumps, one on either side of the main disk, making it look as though Saturn consisted not just of one body, but rather of three. Those two bumps on either side of the main disk were the first ever hint of the rings of Saturn. Unfortunately for Galileo, his telescope wasn't quite good enough to clearly resolve the rings. That would have to wait for the Dutch scientist Christiaan Huygens, in 1655. Still, this was another momentous discovery, and Galileo is often credited, along with Huygens, as the discoverer of the rings.

Eager to claim the credit for his new discovery, Galileo immediately sent out letters to several of his colleagues, including his great colleague and rival, the astronomer Johannes Kepler. Galileo's letter to Kepler (and his other colleagues) was peculiar. Instead of explaining forthrightly what he had seen, Galileo explained that he would describe his latest discovery in the form of an anagram:

smaismrmilmepoetaleumibunenugttauiras

By sending this anagram, Galileo avoided revealing the details of his discovery, but at the same time ensured that if someone else—such as Kepler—later made the same discovery, Galileo could reveal the anagram and claim the credit. This bought him time in which he alone could build upon the discovery. At the same time Galileo also wrote to his patrons, the Medici. But in that letter, eager to keep his patrons happy, Galileo disclosed the full details of his discovery, asking the Medici to keep it secret for the time being. This state of affairs lasted a little over three months, until at the request of Kepler's patron, the Holy Roman Emperor Rudolph II, Galileo relented and revealed that the anagram was the Latin "Altissimum planetam tergeminum observavi," meaning, roughly, that he had observed the highest of the planets (Saturn) to be three-formed.

There is an amusing coda to this story. After Galileo's discovery of the four moons of Jupiter, Kepler developed a theory that Mars must have two moons, on the grounds that Earth had one moon, Jupiter had four, and Mars was the planet between Earth and Jupiter. When Kepler received Galileo's anagram about Saturn he worked hard to decipher it, and finally decoded it as "Salve umbistineum geminatum Martia proles," meaning, roughly, "Be greeted, double knob, children of Mars." Aha, thought Kepler, Galileo must have seen the two moons of Mars! Kepler wasn't sure, though, because one letter in Galileo's anagram went unused. Alas for Kepler, the discovery of the two moons of Mars had to wait until 1877, when far more powerful telescopes were available.

The First Open Science Revolution

Galileo wasn't the only great scientist of the age to use anagrams to announce discoveries. Newton, Huygens, and Hooke all used anagrams or ciphers for similar purposes. In fact, many scientists of the time were reluctant to publicize their discoveries in any way at all. The infamous Newton-Leibniz controversy over who invented calculus occurred in part because Newton claimed to have invented calculus in the 1660s and 1670s, but didn't publish a full account of his discoveries until 1693. In the meantime, Leibniz developed and published his own version of the calculus. Imagine modern biology if publication of the base pairs in the human genome had been delayed by 30 years, or if the base pairs had been announced as an anagram ("AACCGGGT...," say, instead of "CGTCAAGG...")?

Why were Galileo, Newton, and other early scientists so secretive? In fact, a secretive culture of discovery was a natural response to the conditions of the time. There was often little personal gain for scientists in sharing discoveries, and much to lose. Early in his career, Galileo made the mistake of showing a military compass he had invented to a young man named Baldassare Capra. Baldassare later claimed the discovery as his own, and accused Galileo of plagiarism. It took Galileo years of effort and considerable expense to regain the credit for his discovery, not to mention his reputation. No wonder he was so secretive in the matter of Saturn being "three-formed."

Such secretive behavior looks peculiar to our modern eyes. Today, when scientists make a discovery, they share their results as rapidly and as widely as possible, by publishing those results in a scientific journal. For really significant breakthroughs, the scientists involved may write a paper and submit it to a journal in a matter of days. Some scientific journals offer expedited publication services for major papers, promising to publish them within a few weeks after submission. Of course, the reason today's scientists are so eager to share their results is that their livelihoods depend upon it: when a scientist applies for a job, the most important part of the application is their record of published scientific papers. The phrase "publish or perish" has become a cliche in modern science because it succinctly expresses a core fact of scientific life. Modern scientists take this connection between publishing and career success for granted, but in 1610, when Galileo made his string of great discoveries, no such connection existed. It couldn't exist, because the first scientific journals weren't started until 55 years later, in 1665.

What caused this change from a closed, secretive culture of discovery to the modern culture of science, where scientists are eager to publish their best results as quickly as possible? What happened is that the great scientific advances in the seventeenth century motivated wealthy patrons to begin subsidizing science as a profession. This motivation came in part from the public benefit delivered by scientific discoveries, and also in part from the prestige conferred on leaders (such as the Medici) by association with such discoveries. Both motives were best served if discoveries were widely shared through a medium such as the scientific journal. As a result, patrons demanded a shift toward a scientific culture in which it is the sharing of discoveries that is rewarded with jobs and prestige for the discoverer. This transformation was just beginning in the time of Galileo, but two centuries after Galileo's death the culture had changed so much that when the great nineteenth-century physicist Michael Faraday was asked the secret of his success, he replied that it could be summed up in three words: "Work. Finish. Publish." By that time a discovery not published in a scientific journal was not truly complete.

The transformation from a closed, secretive culture of discovery to the more open culture of modern science was one of the most momentous events in history. It resulted in the widespread adoption and growth of the scientific journal system. That system, modest at first, has blossomed into a rich body of shared knowledge for our civilization, a collective long-term memory that is the basis for much of human progress. This system for sharing knowledge has worked tremendously well, and has changed only slowly over the past 300 years.

Today, as we've seen, online tools present a new opportunity, an opportunity to create a collective short-term working memory, a conversational commons for the rapid collaborative development of ideas. At the same time, these tools give us an opportunity to greatly extend and enrich our collective long-term memory. These are tremendously exciting and promising opportunities. We've already seen how open data from projects such as the Sloan Digital Sky Survey is laying the groundwork for a data web that will change the way we explain the world. And we've seen how projects such as Galaxy Zoo, Foldit, and the arXiv are changing the relationship between science and society. But although such examples are encouraging, they fall far short of the potential of networked science. There's a fundamental bottleneck that must be overcome for that potential to be realized. We glimpsed that bottleneck earlier, in the reluctance shown by some scientists to share their data, and in the early lack of interest scientists showed in Wikipedia. Unfortunately, these are not isolated examples, but rather are symptomatic of a more deeply rooted resistance many scientists have to working online. This resistance is holding science back in much the same way that the secretive culture of discovery inhibited science in the seventeenth century. To understand the nature of that resistance, let's take a closer look at some promising-but-failed examples of online tools for scientists.

Science Wikis

Although scientists were reluctant to contribute to Wikipedia in its early days, as Wikipedia has grown, it has inspired several scientists to introduce wikis focused on scientific discovery. An example of such a project is the qwiki (short for "quantum wiki"), set up in 2005 by John Stockton, then a PhD student at the California Institute of Technology (Caltech). Unlike Wikipedia, which aims at a general audience, the qwiki was aimed at professional scientists working in the field of quantum computing. Stockton's goal for the qwiki was to provide a single, centralized reference describing all the latest research in quantum computing and related areas, a sort of rapidly evolving, constantly updated super-textbook. But the qwiki had the potential to go far beyond a textbook: it would be infinitely extensible and modifiable, capable of conveying material ranging from simple introductions of key concepts all the way up to detailed explanations of the latest research results and pointers to unsolved problems at the research frontier. It could include animations and interactive simulations to illustrate key concepts of quantum computing, as well as source materials so other people could further improve those animations and simulations. It could become a locus for Polymath-style collaboration, with theoreticians gathering to attack the deepest theoretical problems of quantum computation, in a new kind of wiki-science. Or experimentalists could gather to share best practices, all the subtle, hard-to-describe details of experiments that often remain tacit knowledge, making it difficult to reproduce results from one laboratory to the next. Even if this vision was only partially realized, the impact on the field of quantum computing would be extraordinary.

The launch of the qwiki was at a workshop I happened to attend, held at Caltech in 2005. The launch caused quite a buzz. In conversation during breaks at the workshop, I heard some people express optimism that the qwiki might do for the specialist knowledge of quantum computing what Wikipedia and Google have done for general knowledge. Unfortunately, that optimism didn't translate into a willingness by those people to contribute. Instead, they hoped someone else would take the lead. After all, why contribute to the qwiki when you could be doing something more useful to your own career, like writing a paper or a grant? Why share your latest and best ideas on the qwiki, when that would only help your competitors? And why contribute to the qwiki when it was still in its beginning stages, and it wasn't yet clear whether it would flourish? The only part of the qwiki that really thrived was the "Researcher pages," vanity pages where individual scientists could add descriptions of themselves and their work. Many scientists were happy to spend an hour or two (and, in some cases, more) fleshing out these vanity pages. But few were willing to spend even ten minutes adding material to other parts of the qwiki. It just wasn't a priority. The result is that today, six years after its launch, the qwiki has failed. Only a few pages of the qwiki are updated with any regularity. Spammers roam the site, adding links to shady products. Nearly all the scientific content on the site was put there by Stockton himself, by people working in the same lab, or by Stockton's successor as maintainer of the qwiki, Stanford University graduate student Anthony Miller. This failure wasn't due to any lack of enthusiasm or capability on Stockton's or Miller's part. They worked hard, adding great quantities of excellent material to the qwiki, and encouraging others to help out. Unfortunately, although many scientists believed such a site had the potential to be a tremendous resource, few were willing to contribute content.

The mindset behind the failure of the qwiki is similar to the mindset I described in the opening chapter of this book, the mindset that makes scientists reluctant to share their data, or to contribute to Wikipedia. At the root of the problem is the monomaniacal intensity that ambitious scientists must bring to the pursuit of scientific publications and grants. For young scientists, especially, this is an intensity borne of the fierce competition for scientific jobs. For example, each year 1,300 people earn physics PhDs from US universities, but only 300 faculty positions in physics open up. At the same time, many PhD programs drum into young scientists the idea that "success" means getting a faculty position at a research-oriented university, and anything else is a failure. The result is a tremendous logjam of scientists trying desperately to get faculty positions. As a young scientist you're not just competing against the other 1,299 newly minted PhDs, you're also competing with people from previous years who are still trying to get faculty jobs. As a result, many young scientists experience great and protracted anguish at their failure to get a faculty job. Even at mid-level universities, a job opening can easily draw more than 100 applicants. In such a competitive environment, 80-plus-hour workweeks are common, and as much time as possible is devoted to the goal that will result in a position at a top university: an impressive record of scientific papers. The papers also bring in the research grants and letters of recommendation necessary to be hired. Scientists who already have tenured positions continue to need grant support, which requires a strong work ethic still focused on producing papers. Given all this, how could a scientist possibly have the time to contribute to efforts such as the qwiki? They may agree in principle that they'd like the qwiki to succeed, but in practice they're too busy writing papers and grant proposals to have any time to contribute themselves.

The qwiki is just one of many science wikis that have been launched. Similar efforts have been made to develop wikis for genetics, string theory, chemistry, and many other subjects. Like the qwiki, many of these science wikis had great potential, and some generated considerable buzz and optimism in their fields. But most have failed to take off, foundering beneath scientists' lack of time and motivation to contribute. Those science wikis that do succeed are usually in a supporting role for some more conventional project. Many laboratories, for example, run internal wikis as a way of storing reference materials for their experiments. Another successful wiki comes from the Polymath Project, which uses its wiki as a place to distill the most valuable insights from the Polymath collaboration. The Polymath wiki has attracted many thousands of edits, by more than 100 users, and at peak times attracts dozens of edits and thousands of pageviews per day. (Note that I set up the Polymath wiki, and am not an independent judge of its success.) Again, though, the Polymath wiki is in support of a conventional goal: solving a mathematical problem and writing a paper. In each of these cases, the wiki has not been an end in itself. Wiki-science, as promising as it might be, remains a dream.

User-Contributed Comment Sites for Science

It's not just science wikis that are failing. Several organizations have created user-contributed comment sites where scientists can share their opinions of scientific papers, and so help other scientists decide which papers are worth reading, and which aren't worth the effort. The idea is similar to sites such as Amazon.com, which collect customer reviews of books, electronic gadgets, and other products. As anyone who's ever used Amazon.com knows, the reviews can be very helpful when deciding whether to buy a product. Maybe something similar would be helpful for scientists?

The user-contributed comment site with the highest profile was created by one of the most prestigious publishers in science, *Nature*. In 2006, *Nature* launched a site where scientists could write open comments on papers that had been submitted to *Nature*. Despite much effort and publicity, the trial was not a success. The final report terminating the trial explained:

> There was a significant level of expressed interest in open peer review. . . . A small majority of those authors who did participate [in the trial] received comments, but typically very few, despite significant web traffic. Most comments were not technically substantive. Feedback suggests that there is a marked reluctance among researchers to offer open comments.

In other words, while lots of people wanted to read comments about other people's papers, almost no one wanted to actually write comments.

The *Nature* trial is just one of many attempts to build user-contributed comment sites for science. Physics, in particular, has seen many such sites, perhaps because it was the first field to broadly adopt the web as a way of distributing scientific papers. The first attempt was the site Quick Reviews, which came online in 1997, and was discontinued for lack of use in 1998. A similar site, Physics Comments, was built a few years later, but suffered the same fate, being discontinued in 2006. A still more recent site, Science Advisor, is still active, but has more members (1,240) than reviews (1,119) as I write. It seems that many scientists want to read comments on scientific papers, but very few want to volunteer to write such comments.

Why are the user-contributed comment sites failing? In principle, most scientists agree that it would be tremendously useful if thoughtful commentary on scientific papers was widely available. But if that's true, then it seems like a puzzle that these sites—many of them well designed and well supported—fail, when the comment sections on sites such as Amazon.com thrive. The problem the scientific comment sites have is that while thoughtful commentary on scientific papers is tremendously useful for *other* scientists, that doesn't mean it's in anyone's individual best interest to write comments. Imagine how things look from the point of view of an individual scientist considering commenting on such a site. Why write a comment when you could be doing something more useful to you individually, like writing a paper or a grant? Even if you did write a comment, you'd likely be reluctant to publicly criticize someone else's paper. After all, the person you criticize might be an anonymous referee in a position to scuttle your next paper or grant application.

The contrast between the failures of the user-contributed comment sites for science and the success of the Amazon.com reviews is stark. To pick just one example, you'll find more than 1,500 reviews of Pokemon products at Amazon.com, more than the total number of reviews on all the science comment sites I described above. You may object that there are more people who buy Pokemon products than there are scientists. That's true. But there are still more than a million professional scientists in the world, and those scientists spend much of their working lives forming opinions of papers written by others, far more time than even the most enthusiastic parents can spend on Pokemon. It's a ludicrous situation: popular culture is open enough that people feel a desire to write Pokemon reviews, yet scientific culture is so closed that scientists won't publicly share their opinions of scientific papers in an analogous way. Some people find this contrast curious or amusing; I believe it signifies something seriously amiss with science.

The Modern Challenge for Open Science

The failure of the science wikis and the user-contributed comment sites for science is part of a much larger pattern. Projects such as the Polymath Project, Galaxy Zoo, and Foldit have all been very successful, but that success has come in part because of a fundamental conservatism: all of them ultimately aim to produce scientific papers. Tools such as the science wikis and user-contributed comment sites break away from this conservatism, since contributions to such sites are ends in themselves, and don't directly result in scientific papers. Unfortunately, the result is that career-minded scientists have little incentive to contribute to such sites, and instead focus their efforts on doing what is rewarded: writing papers. The grand ideas for amplifying collective intelligence that we discussed in part 1 have little chance to thrive when incremental ideas such as science wikis and user-contributed comment sites are already beyond the pale. Many of the tools with the potential to most dramatically change and improve how science is done are simply nonstarters. It's no accident that so many of the best examples of amplifying collective intelligence in part 1 came from outside science; too often scientists are lagging, not leading in the development of new tools for the production of knowledge. And although we have seen some impressive science-oriented projects, they explore only a tiny fraction of the landscape of possibilities. We're missing a giant opportunity.

Indeed, even the possibilities that are being explored are not thriving as they should. While undertakings such as the Sloan Digital Sky Survey and the Human Genome Project are opening up their data to other scientists, the data from the great majority of scientific experiments remains closed. Scientists typically have little incentive to disclose their data, and so instead they hoard it. In the words of medical researchers Elizabeth Pisani and Carla AbouZahr, in science it's "publish [papers] or perish," not "publish [data] or perish." And as long as that remains true, much of the world's scientific knowledge will remain locked up, preventing the scientific data web from reaching its full potential.

Equally concerning are the disincentives for scientists to develop new online tools. While I was writing this book, a well-known physicist told me that Paul Ginsparg, the physicist who created the arXiv, had "wasted his talent" for physics by creating the arXiv, and that what Ginsparg was doing was like "garbage collecting": it was good that someone was doing it, but beneath someone of Ginsparg's abilities. Keep in mind that this astonishing narrow-mindedness was coming from a person who uses the arXiv every day. Ginsparg has perhaps done more for physics (not to mention the rest of humanity) than any other physicist of his generation. Yet sentiments such as these are often voiced privately by scientists. People who build tools such as the arXiv are dismissed as "mere" tool builders, as though it is somehow unworthy to be building tools that speed up the whole process of doing science. This lack of regard extends to the institutional level, where there is often little support for building new tools. Projects such as Galaxy Zoo and the arXiv often begin with little or no funding, in part because their first stage involves creating a tool, not writing a paper. How can ideas such as citizen science and the data web reach their potential in an environment where building new tools is held in such low regard?

The overall pattern, then, is that networked science is being strongly inhibited by a closed scientific culture that chiefly values contributions in the form of scientific papers. Knowledge shared in nonstandard media isn't valued by scientists, regardless of its intrinsic scientific value, and so scientists are reluctant to work in such media. The potential of networked science—ideas such as the data web, citizen science, and collaboration markets—is thus remaining unrealized. To reach its full potential, networked science must be open science.

The irony in all this is that the value of openly sharing scientific information was deeply understood by the founders of modern science centuries ago. It was this understanding that led to the modern journal system, a system that is perhaps the most open system for the transmission of knowledge that could be built with seventeenth-century media. The adoption of that system was achieved by subsidizing scientists who published their discoveries in journals. But that same subsidy now *inhibits* the adoption of more effective technologies, because it continues to incentivize scientists to share their work in conventional journals, while there is little or no incentive for them to use or develop modern tools. Indeed, when the scientists of today resist sharing their data and ideas, they are unconsciously echoing the behavior of Galileo, Newton, and company, with their secrecy and their anagrams. It may be a practical response to immediate personal concerns, but over the long run it's the wrong way to do science.

To take full advantage of modern tools for the production of knowledge, we need to create an open scientific culture where as much information as possible is moved out of people's heads and laboratories, and onto the network. This doesn't just mean the information conventionally shared in scientific papers, but *all* information of scientific value, from raw experimental data and computer code to all the questions, ideas, folk knowledge, and speculations that are currently locked up inside the heads of individual scientists. Information not on the network can't do any good.

In an ideal world, we'd achieve a kind of *extreme openness*. That means expressing all our scientific knowledge in forms that are not just human-readable, but also machine-readable, as part of a data web, so computers can help us find meaning in our collective knowledge. It means opening the scientific community up to the rest of society, in a two-way exchange of information and ideas. It means an ethic of sharing, in which all information of scientific value is put on the network. And it means allowing more creative reuse and modification of existing work. Such extreme openness is the ultimate expression of the idea that others should be able to build upon and extend the work of individual scientists, perhaps in ways they themselves would never have conceived. In practice, there will need to be some limits—think of concerns such as patient confidentiality in medical research—and we'll discuss those limits in the next chapter. But even within those limits, the openness I am advocating would be a giant cultural shift in how science is done, a second open science revolution extending and completing the first open science revolution, of the seventeenth and eighteenth centuries. In the next chapter, we'll discuss how this more open culture can be achieved.

An Aside on Commercialization and Secrecy in Science

In this chapter, we've seen how scientists' strong commitment to papers as the ultimate expression of scientific discovery is inhibiting new and better ways of doing science. But for some scientists there's an additional inhibition, and that's a need for secrecy because they're pursuing patents and commercial spin-offs from their work. As an example, from 2001 to 2003 I was part of a large research center working on quantum computing. Although the center was a long way from producing a commercial product, the center's leaders hoped that one day there would be such spin-offs. When scientists attended research seminars at the center, they were (for a while) asked to sign nondisclosure agreements promising not to talk with other people about the content of the seminars. Many scientists at the center meticulously documented their work in notebooks where each page was dated and signed by center officials, to help establish priority in the event of later patent applications. Such secretiveness may help lead to commercial success. But it's impossible for such a culture to coexist with the open collaborative atmosphere that is seen in, for example, the Polymath Project, or that is required for wiki-science to succeed.

Such commercially driven secrecy is relatively new in our universities, where most basic research is done. Indeed, until quite recently, universities focused most of their scientific research effort on basic research without immediate commercial application. This has changed over the past few decades, in large part because of a piece of legislation called the Bayh-Dole Act, passed by the US Congress in 1980. What Bayh-Dole did was to give US universities (rather than the government, as was formerly the case) ownership of patents and other intellectual property produced with the aid of government grants. After Bayh-Dole passed, many universities began to broaden their focus beyond basic research, supporting more applied research in the hope of making money from commercial spin-offs. Simultaneously, and for the same reason, there was also an increase in patents related to the basic research conducted at universities. Many other countries have followed the US lead and passed legislation similar to Bayh-Dole, with a similar effect on their research culture. The success of these efforts is questionable—many universities actually lose money attempting to commercialize their research—but interest in commercialization and intellectual property has nonetheless made many scientists more secretive.

This commercially driven secrecy is a big cultural shift in our universities. Historically, before Bayh-Dole and similar legislation, the results of basic science were usually (eventually) openly disclosed, in the form of papers, in the belief that an improved understanding of how the world works would benefit everyone over the long run. For instance, basic research on electricity and magnetism was the foundation for inventions such as motors and electric lighting and radio and television. Basic research on quantum mechanics was crucial for the semiconductor industry. It's the familiar idea that a rising tide floats all boats. And so there was a fairly clean split in our innovation system. On one side was the basic research system, whose ultimate results were shared publicly as research papers, on the grounds that over the long term everyone would gain. On the other side was privately funded applied research that aimed at short-term product development, and that was often carried out in secret. Bayh-Dole has begun to break this division down, and today governments and grant agencies increasingly see the pursuit of patents and other intellectual property as a major reason to support basic research.

This change is a genuine impediment to the open sharing necessary for networked science to thrive. However, we should keep the size and scope of this impediment in proper perspective. While writing this book, I sometimes spoke with people who assumed that commercially driven secrecy is the single biggest obstacle to open science. That is incorrect. In large parts of basic science, scientists' concerns about commercialization are decidedly secondary compared to their relentless focus on conventional publication. Commercialization and patent rights are welcome if they come, but career success comes by earning the esteem of peers through publication. This is most evident in job applications:

scientists often list a few patents or spin-offs resulting from their work, but the emphasis is on papers, papers, papers, and grants, grants, grants. This is true in large parts of physics and astronomy, in mathematics, in substantial parts of chemistry, biology, and in many other fields of science. In these fields, the immediate obstacle to open science isn't commercialization, it's a culture that only values and rewards the sharing of scientific knowledge in the form of papers.

In a few areas of basic science, commercially driven secrecy is paramount. This is true in some of the early-stage work that may lead to later drug development, for instance. In such fields, science will likely remain a closed, secretive affair. And there is a much larger gray area in basic science where concerns about commercial secrecy are a factor, but not always a dominant factor. The real problem is scientific work that could in principle be open, but where unfounded hopes of later patents impede open science. Over the long run, there is a conversation to be had about the role of intellectual property in basic science. But the foundation for open science, the place where we should start, is with a change in the culture of science so that it doesn't just value and reward the writing of papers, but also new ways of sharing. That's the most crucial problem, and it's to that problem we now turn.

CHAPTER 9

The Open Science Imperative

Imagine you're a working scientist who believes wholeheartedly that open science will bring enormous benefits to science and to our society. You understand that changing the deeply entrenched culture of science will be difficult, but decide nonetheless to go all out sharing your ideas and data online, contributing to new tools such as science wikis and user-contributed comment sites, and making the code for your computer programs freely available. All this takes a great deal of time and effort, and yet you find that without colleagues willing to reciprocate, the benefits to you are small. That's because many of the benefits of open science only come if it is collectively adopted by large numbers of scientists. And as just one scientist you can't compel everyone else to do open science.

A typical experience is that of my colleague and former student Tobias Osborne, now of the University of Hannover in Germany. Eager to try out open science, for six months Osborne carried out much of his research on quantum computing in the open, on a blog. He wrote many thoughtful posts, full of insightful ideas, and his blog attracted a following in the quantum computing community, with more than 50 regular readers. Unfortunately, few of those readers were willing to provide much feedback on Osborne's posts, or to share their own ideas. And without a community of engaged colleagues, it was a lot of effort to work in the open, for only a small return. Osborne ultimately concluded that open science won't succeed because it "would require most scientists to simultaneously and completely change their behaviour." Experiences such as this make open science seem like a hopeless cause.

Although it's true that it will be difficult to move toward open science through direct action by individual scientists, that doesn't mean other approaches can't succeed. Our society has solved many problems analogous to the open science problem, problems where direct individual action doesn't work, and benefits only come if many people in a large group simultaneously adopt a new way of doing things. An example is the problem of which side of the road to drive on. If you live in a country where people drive on the left, you can't one day start a movement to drive on the right merely by swapping the side you personally drive on. But that doesn't mean it's not possible for everyone to switch simultaneously. That's exactly what happened in Sweden on September 3, 1967, at 5 am. There were good reasons for the Swedes to switch: the people in neighboring countries already drove on the right, and in addition, most vehicles in Sweden were already left-hand drive, making driving on the right actually safer. But, as with open science, the mere fact that driving on the right would be better wasn't enough to cause a change through direct action by individuals. Instead, it required an extended campaign by the government, and a change in the law.

Changing sides of the road seems far removed from changing the culture of science. But, in fact, the first open science revolution required a similar type of collective action. We've seen how seventeenth century scientists often kept their results to themselves—unless you count sending anagrams as sharing! When the scientific journal system was first introduced, many scientists were suspicious, unwilling to share their results with others in a new medium. While individual scientists could see that science as a whole would progress more quickly if *all* scientists shared news of their discoveries freely, that didn't mean it was in their individual best interest to publish in the journals. This posed a problem for the editors of early journals, people such as Henry Oldenburg, who founded the world's first scientific journal, the *Philosophical Transactions of the Royal Society*, in 1665. Oldenburg's biographer, Mary Boas Hall, tells of how Oldenburg would write to the scientists of the day and "beg for information," sometimes writing simultaneously to two competing scientists on the grounds that it would be "best to tell A what B was doing and vice versa, in the hope of stimulating both men to more work and more openness." In this way, Oldenburg provoked some of the most eminent scientists of his day, including Newton, Huygens, and Hooke, to publish in the *Philosophical Transactions*. The need for such subterfuge ceased only after decades of work by Oldenburg and others to change the culture of science.

The common pattern underlying the problem of switching sides of the road and the problem of open science—both today and in the seventeenth century—is that the interests of individuals aren't naturally aligned with the collective group interest. Someone who believes "everyone should do this" e.g., open science or switching sides of the road, doesn't necessarily also believe "I should do this, even if no one else does." Social scientists call problems like this collective action problems. The trick to solving collective action problems is to figure out ways of aligning individual interest with the collective interest. In the case of Sweden's switch to the right, the solution was to use the government's legitimacy—expressed, in part, through the force of law—to compel people to switch. One day it was in people's individual best interest to drive on the left, while the next day it was in their interest to drive on the right. Similarly, the genius of the first open science revolution was to align individual and collective interest by rewarding scientists for sharing their discoveries in scientific journals. The problem today is that while it's now in the collective interest for scientists to adopt new technologies, their individual interests remain aligned with journal publication. We need to bring the individual interest back into alignment with the collective interest.

The good news is that a lot is known about how to solve collective action problems. Writing in the 1960s, the political economist Mancur Olson analyzed what he called the "logic of collective action," trying to understand the conditions under which individuals in a group will work together in their collective interest, and those under which they will not. In the 1990s, the political economist Elinor Ostrom substantially deepened Olson's analysis for a particular type of collective action, namely, how groups can work together to manage resources that they hold in common, such as water and forests. The books Olson and Ostrom wrote describing their work are among the most frequently cited books ever in the social sciences, and the work has been so influential that Ostrom was awarded the 2009 Nobel Prize in Economics.

I mention this work as an antidote to pessimism about open science. Some very clever people have spent a great deal of time investigating real-world examples where collective action problems have been solved, and have thought hard about how the strategies used in those examples can be generalized to solve other collective action problems. What Olson, Ostrom, and their colleagues have shown is that while solving collective action problems is difficult, it's not impossible. Before we give up on open science, we should draw on these ideas. We'll now look at two strategies that can be used to shift the culture of science. Neither strategy is a quick fix, but with enough imagination and determination these strategies can make science far more open. Although my account is based on the work of Olson and Ostrom and their successors, I won't make the connections explicitly, since this isn't a textbook on political economy. If you're interested in exploring the connections further, please see "Selected Sources and Suggestions for Further Reading," beginning on page 217.

Compelling Open Science

Earlier in this book we discussed the open access policies that some scientific grant agencies are introducing, in order to make the results of scientific research broadly available. Recall, for example, that the US National Institutes of Health (NIH) now requires scientists to make their papers openly accessible within 12 months of publication. Scientists who don't agree to this condition must look elsewhere for funding. It's a policy of compulsion, similar to the strategy used by the Swedish government to switch sides of the road. In this way, powerful organizations such as governments and grant agencies can cause everyone in a community to simultaneously change their behavior.

Following on from their open access policies, several grant agencies now require scientists to openly share their *data*. These open data policies are in the spirit of the Bermuda Agreement to share human genetic data (see page 7), but broader in scope. There are a lot of ways this is happening; so let me describe just a few snapshots. In narrowly focused areas such as genomics, the policies can be quite demanding. Earlier in the book, on page 3, we saw how genomics can be used to figure out links between genes and disease; the resulting studies are called genome-wide association studies (GWAS). In 2007 the NIH instituted a policy requiring that data from GWAS be made openly available, subject to some restrictions to ensure participant privacy. Another major funder of genomics research, the Wellcome Trust, now requires all genetic data to be made openly available, again subject to privacy and similar concerns. Furthermore, these agencies specify which online databases the data should be uploaded to, in what formats, and so on.

Broader policies on data sharing are usually less specific. For instance, since 2006 the UK Medical Research Council has required all scientists it funds to make their data openly available, provided that doesn't violate any ethical or legal regulations. But this policy doesn't specify exactly how or where data should be made available. Many open data policies are still in early stages of development. For instance, since January of 2011 the US National Science Foundation has required grant applications to include a two-page data management plan. It's not a full-fledged open data policy, but a spokesperson said this announcement was merely "phase one" of an effort to ensure that all data be openly accessible. Overarching all this, at the highest political levels there is a growing understanding of the value of open data. For instance, in 2007 the Organization for Economic Co-operation and Development (OECD) recommended that member countries make publicly funded research data openly accessible. Such recommendations take time to filter down, but over time they can have an impact.

Open access and open data policies are powerful steps toward open science, the sorts of steps that are difficult for individual scientists to take on their own. The grant agencies are the de facto governance mechanism in the republic of science, and have great power to compel change, more power even than superstar scientists such as Nobel prizewinners. The behavior of many scientists is dictated by the golden rule: them that have the gold make the rules. And the big grant agencies have the gold. If the people running the grant agencies decided that as part of the granting process, grant applicants would have to dance a jig downtown, the world's streets would soon be filled with dancing professors. Now, many people— including many grant officers—find fault with this system, believing that it is too centralized and controlling. But as a practical matter, the grant system presently rules much of science, and if the grant agencies decide to take open science seriously, so too will scientists. Imagine, for example, that one of the big grant agencies began asking applicants to submit evidence of public outreach using blogging and online videos. Or suppose they started asking applicants to describe their contributions to science wikis, as evidence of research activity. Such policies would do much to legitimize new tools.

Although grant agencies can help new tools become accepted, they don't have unlimited power to impose open science on scientists. Recall again the story of the Bermuda Agreement for the sharing of human genetic data. Those principles weren't merely imposed by fiat on molecular biologists by some central granting agency. Instead, leaders in the molecular biology community gathered in Bermuda, where they agreed that it would be in the whole community's best interest to share data. Essentially, individual scientists were saying, "We'd like to go open—but only if everyone else does too." The granting agencies then helped achieve that end by enforcing the policy of openness. But part of the reason the policy was so effective was because it already had the support of leading molecular biologists. A similar situation occurred in Sweden, where the switch to the right-hand side of the road was only made after a decades-long public discussion of the idea.

To be successful, grant agencies can't merely compel openness, they must also forge consent and agreement within the scientific community. If they don't do this, it's too easy for scientists to respond by following the letter of grant agency requirements, but not the spirit. Imagine future scientists releasing "open" data sets that are so poorly documented that they're useless to anyone else. It's one thing for a scientist to dump raw data online in some obscure location. It's quite another to carefully document and calibrate that data, to integrate it with other scientists' data, and to actively encourage other scientists to find new uses for it. That's what it will take for the scientific data web to succeed. More generally, for networked science to reach its full potential, scientists must make an enthusiastic, wholehearted commitment to new ways of sharing knowledge. For that to happen, grant agencies must work individually with scientific communities, talking at length with scientists in each community about ways that community could become more open. Are there data that could be systematically shared? What about computer code? What about people's questions and ideas and folk wisdom? What else could be shared? How quickly could it be shared? What new tools need to be developed to make this effective? If the grant agencies do this, they can act as catalysts for Bermuda-style agreements to share scientific knowledge. And, having forged such agreements, they can then express them in policy. This will be long, slow work, but the payoff will be a tremendous cultural shift toward more openness.

Incentivizing Open Science

The prospect of the grant agencies saying "Thou shalt work more openly" leaves me, as a scientist, with mixed feelings. While it will promote the use of new tools, it won't cause truly enthusiastic adoption of those tools by scientists, unless we also create new incentives to use those tools. Today's scientists show a relentless drive to write papers because that's what's valued by the scientific community. We need new incentives that create a similar drive to share data, code, and other knowledge. How can we make sharing knowledge in new ways just as imperative for scientists as publishing papers is today?

It helps to look at this question in economic terms. In a conventional economy, if I trade you a sofa in return for some cash, you gain a sofa, and I lose a sofa. But scientific discoveries are different. If I share news of a discovery with you, I don't lose my knowledge of that discovery. This kind of sharing is great for society as a whole, but it has a problem from the point of view of the original discoverer: if they are not recompensed, they have much less reason to invest time and effort to come up with the discovery in the first place.

The solution to this problem adopted by the scientific community in the seventeenth century (and still used today) is brilliant. Instead of giving people exclusive rights to their ideas, as in a conventional economy, we have created an economy based on reputation. Scientists openly share their discoveries by publishing them in scientific papers—essentially, giving them away—but in return they get the right to be credited as the discoverer. By being so credited they can build up a reputation, which can be turned into a paying job. It's a type of property rights in ideas, leading to an economy based on reputation, and establishing an invisible hand for science that strongly motivates scientists to share their results. The foundation for this reputation economy is a set of very strong social norms: scientists must credit other people's work; they cannot plagiarize; and scientists judge other scientists' work by their record of published papers. But these norms focus on just one way of sharing scientific knowledge: the scientific paper. If we could establish similar norms and a reputation economy that encourages broader sharing of scientific knowledge, then the invisible hand of science would become stronger, and the process of science would be greatly accelerated.

How can we expand science's reputation economy in this way? Let's look at an example where such an expansion is beginning to happen today. It's a story that involves both the arXiv—the service we saw earlier that makes the latest results of physics available for free download—and another service for physicists called SPIRES. The arXiv and SPIRES are together creating incentives for physicists to share knowledge in new ways. To explain what's going on, I first need to explain what SPIRES does. Suppose that, for some reason, you're very interested in finding out what impact Stephen Hawking's latest arXiv preprint is having on other scientists' work. SPIRES can help by telling you which arXiv preprints and published journal papers are citing Hawking's preprint. SPIRES might tell you, for example, that not a single preprint or paper has yet cited Hawking's latest. Or maybe you'll find that it's spurred many other physicists to work on related ideas. SPIRES can also give you the big picture of how often Hawking's (or any other physicist's) preprints and papers have been cited in aggregate, and who is citing them. This makes SPIRES a tremendously useful tool for evaluating candidates for scientific jobs. When physics hiring committees meet to evaluate candidates in the areas that SPIRES covers (particle physics and some related areas), it's not unusual for everyone in the meeting to have their laptops out, comparing SPIRES citation records.

What's all this got to do with openness and new incentives to share knowledge? Well, a couple of decades ago, preprints were viewed by most physicists as mere stepping-stones along the road to conventional journal publication. They weren't valued as ends in themselves. To build your career, you needed a record of high-quality journal papers. Today, because of the arXiv and SPIRES, preprints have some status as ends in themselves. It's not uncommon for physicists to, for example, list preprints that have not yet been published in a journal on their curriculum vitae. And if a physicist discovers someone else working on a project that competes with one of their own projects, they may rush to get their preprint out first. Preprints don't yet have as high a status as journal articles, but a preprint with hundreds of SPIRES citations can still carry quite a punch, career-wise. By providing a way of demonstrating the scientific value and impact of a preprint, SPIRES and the arXiv have created a real incentive for physicists to produce preprints, an incentive that's separate from the usual incentive to write papers.

I've got to admit that as cultural changes go, this one's pretty small. The move to a preprint culture in physics does speed up the sharing of scientific knowledge, and makes that knowledge more broadly accessible. But it's not nearly as big a change as replacing anagrams by scientific journals! Still, we should pay attention to the story of the arXiv and SPIRES, because it shows that it really is possible to create new incentives for scientists to share knowledge. What's more, this happened without any compulsion by a central agency. Once SPIRES enabled the impact of preprints to be mea-sured, the new incentive emerged naturally as individual physicists started using the SPIRES citation reports. In science, as in so many parts of life, what gets measured is what gets rewarded, and what gets rewarded is what gets done.

Could a similar strategy be used to incentivize scientists to share other types of scientific knowledge? Let's think, for example, about incentives to share data. Suppose that, as has happened with preprints in physics, scientists began to regularly cite other people's data in their own scientific papers. This is already starting to happen, and will happen more as open data policies become more common. And suppose someone sets up a citation tracking service that not only tracks citations to papers and preprints, but also citations to data. If the service is good, people will use it to assess other scientists. And they'll start to see more vividly the impact data sharing has. At this point, sharing data will start to help rather than hurt scientists' careers. Indeed, not only will scientists have an incentive to share their data, it will be to their advantage to make that data as useful as possible to other scientists. Scientists will begin to see building the data web as an important part of their job, not as a distraction from the serious business of writing papers.

This same kind of incentive building can be applied to any type of scientific knowledge: preprints, data, computer code, science wikis, collaboration markets, you name it. In each case the overall pattern is the same: citation leads to measurement leads to reward leads to people who are motivated to contribute. This is a way of expanding science's reputation economy. There will, in practice, be many complications, and many possible variations on this theme. Indeed, even the arXiv-SPIRES story I told was oversimplified: SPIRES was just one factor among several that gave preprints real status in physics. But the basic picture is clear.

A case of particular importance is computer code. Today, scientists who write and release code often get little recognition for their work. Someone who has created a terrific open source software program that's used by thousands of other scientists is likely to get little credit from peers. "It's just software" is the response many scientists have to such work. From a career point of view, the author of the code would have been better off spending their time writing a few minor papers that no one reads. This is crazy: a lot of scientific knowledge is far better expressed as code than in the form of a scientific paper. But today, that knowledge often either remains hidden, or else is shoehorned into papers, because there's no incentive to do otherwise. But if we got a citation-measurement-reward cycle going for code, then writing and sharing code would start to help rather than hurt scientists' careers. This would have many positive consequences, but it would have one particularly crucial consequence: it would give scientists a strong motivation to create new tools for doing science. Scientists would be rewarded for developing tools such as Galaxy Zoo, Foldit, the arXiv, and so on. And if that happened we'd see scientists become leaders, not laggards, in developing new tools for the construction of knowledge.

There are limits to the citation-measurement-reward idea. Obviously, it's neither possible nor desirable to judge a discovery based solely on what citations a paper (or preprint or data or code) has received. When it comes to assessing the importance of a discovery, there's no replacement for understanding the discovery deeply. But with that said, the basis for the reputation economy in science is the citation system. It's the way scientists track the provenance of scientific knowledge. If scientists are to take seriously contributions outside the old paper-based forms, then we should extend the citation system, creating new tools and norms for citation, while keeping in mind the limitations citations have (and have always had) as a way of assessing scientific work.

Today, many scientists find the idea of working more openly almost unimaginable. After giving talks about open science I've sometimes been approached by skeptics who say, "Why would I help out my competitors by sharing ideas and data on these new websites? Isn't that just inviting other people to steal my data, or to scoop me? Only someone naive could think this will ever be widespread." As things currently stand, there's a lot of truth to this point of view. But it's also important to understand its limits. What these skeptics forget is that they *already* freely share their ideas and discoveries, whenever they publish papers describing their own scientific work. They're so stuck inside the citation-measurement-reward system for papers that they view it as a natural law, and forget that it's socially constructed. It's an agreement. And because it's a social agreement, that agreement can be changed. All that's needed for open science to succeed is for the sharing of scientific knowledge in new media to carry the same kind of cachet that papers do today. At that point the reputational reward of sharing knowledge in new ways will exceed the benefits of keeping that knowledge hidden. Now, at this point skeptics will sometimes say, "But no one will ever take ideas shared on a *blog* [or wiki, etc.] seriously!" This may be true right now—although even that is changing—but over the long run, the view is myopic and ignores the lessons of the first open science revolution. We have a real chance to drive the same kind of transition that Henry Oldenburg and his colleagues caused in the seventeenth and eighteenth centuries, incentivizing scientists to share their scientific knowledge using the most powerful tools available today. We can bring the interests of individual scientists back into alignment with the collective interest of the scientific community and the public as a whole: driving science forward as fast as possible.

Limits to Openness

What limits should be imposed on openness in science? Although it's broadly true that, as I said earler, information not on the network can't do any good, some limits are necessary. Some of these limits are obvious: doctors can't share patient records willy-nilly, security experts can't share information that compromises security, and so on. Of course, there are already many measures in place to prevent disclosure of information when it would violate expectations of privacy, ethics, safety, and legality. But there are more subtle concerns about openness that also need to be considered.

Might openness overwhelm scientists? One of the great mathematicians of all time, Alexander Grothendieck, believes that it was his capacity to be alone that was the wellspring of his creativity. In autobiographical notes, he says that he found true creativity as a consequence of being willing to "reach out in my own way to the things I wished to learn, rather than relying on the notions of the consensus, overt or tacit, coming from a more or less extended clan of which I found myself a member." Grothendieck is not alone in this belief. Ideas that require careful nurturing may wither and die if they are modified prematurely in response to others' opinions. Perhaps if we move to a more open, collaborative culture, we risk giving up the independence of mind necessary for the highest forms of creativity. Will fewer people attempt bold work that does not fit within the shared praxis of an existing scientific community, but which instead aims to define a new praxis?

There's a general problem here that goes beyond Grothendieck's desire for solitude, or romantic notions of lone geniuses redefining fields. It's the problem, which we discussed at the end of chapter 3, that scientists only have limited time, and this imposes constraints on how they work with others. Should they collaborate a little, a lot, or not at all? If they choose to collaborate, with whom should they work? No matter how much they enjoy collaboration, their attention doesn't scale infinitely, and so must be managed carefully. Sometimes the resolution of the problem is, as for Grothendieck, to seek solitude. But for scientists who choose to collaborate, the problem manifests in other ways. In the Polymath Project, for example, a small number of contributions came from people without the mathematical background to make significant progress on the problem. Those people were outside the praxis shared by most Polymath participants. Although their contributions were well intentioned, they were of little help. Fortunately, there were few low-quality contributions, and they were easily ignored. But if there had been more, they would have significantly taxed the attention of other Polymath participants. Similar problems can be caused by cranks, trolls, and spammers, or even people who are just plain unpleasant.

These problems are serious but not insurmountable. A system can be open without requiring that all participants receive equal attention. And you can share your knowledge openly, without having to pay attention to everyone (or, indeed, anyone) else. In general, for open collaborative systems to work most effectively, participants must have powerful ways of filtering information, so they can concentrate on the information of most interest to them, and ignore the rest. In the MathWorks competition, for example, recall how the score helps participants filter out unhelpful ideas, and focus on the best ideas from other users. And if low-quality contributions become more of a problem in the Polymath Project, it too could be filtered. Ideally, science is open-but-strongly-filtered. This is a natural consequence of the fact that while our attention doesn't scale, sharing knowledge does. In an open-but-filtered world there is no problem with people such as Grothendieck pursuing their own solitary program.

Won't open science sometimes be used for ends that many scientists find distasteful? In November of 2009, hackers broke into a computer system in one of the world's leading centers for climate research, the Climate Research Unit at the University of East Anglia, in the UK. The hackers downloaded more than 1,000 email messages sent between climate scientists. They then leaked the emails (and many other documents) to bloggers and journalists. The incident received worldwide media attention, as many climate change skeptics seized upon the emails, claiming that they contained evidence to prove that the notion of human-caused climate change was a conspiracy among climate scientists. One of the examples used to support this claim was an email from Kevin Trenberth, a well-known climate scientist from the National Center for Atmosphere

Research in Boulder, Colorado. In his email, Trenberth says; "The fact is that we can't account for the lack of warming at the moment and it is a travesty that we can't." In fact, the sentence was being quoted badly out of context. In the email, Trenberth was discussing a paper he'd recently published, which was looking at the causes of the year-to-year variation in the Earth's surface temperature—why we have hotter and colder years—and how that variation relates to the long-term overall increase in temperature. The year-to-year variations are presumably due to changes in the way surface heat is redistributed into the ocean, into melting ice, and so on. Trenberth's email and paper were pointing out that we don't fully understand all the processes causing these variations, and so we can't necessarily explain why any given year is hotter or colder. Although the email expressed some frustration at this state of affairs, it didn't in any way contradict his belief in the long-run rise in temperature, which swamps the short-term variations. Note that the issue here is not about whether you agree with Trenberth about climate change. The issue is that a careful and honest skeptic of climate change could not possibly interpret Trenberth's email as expressing any doubt on his part that humans are causing climate change. Nevertheless, many skeptics chose to quote the sentence out of context, either maliciously, to further their own ends, or carelessly, from genuine ignorance of the original intent.

This kind of incident illustrates a major risk facing climate scientists who are considering working more openly. On the one hand, open sharing of ideas and data has the potential to speed up discovery. On the other hand, every piece of information shared by climate scientists, no matter how innocent, stands a chance of being attacked by groups who want to bring climate science into disrepute by exaggerating minor problems, or by reporting remarks like Trenberth's out of context. Given this, how openly should work on climate science be done? This is not an easy question to answer. If the issues were solely scientific, then the climate scientists should move quickly to work more openly. But the issues aren't just scientific, they're also political. I believe that the right approach is not to make a dramatic shift, but rather to move gradually toward a more open system, diagnosing and fixing problems as they arise.

Might open science lead to the spread of misinformation?
Over the past two decades scientists have discovered more than 500
planets orbiting around stars other than our sun. These discoveries
are exciting, but until recently, all the confirmed extrasolar planets
were gas giants, more like Jupiter or Neptune than they are like the
Earth. Hoping to change this situation, in early 2009 NASA launched
the Kepler Mission, a space-based observatory that astronomers
believed could discover the first Earth-size planets orbiting around
other stars. NASA policy ordinarily requires open release of data
from such missions within a year, and it was widely expected by
scientists that the Kepler data would be released in June of 2010.
But in April of 2010 a NASA advisory panel granted an unusual
extension, allowing the Kepler team to withhold data on the 400 best
planet candidates until February of 2011. That gave them more time
to analyze the data, and a better shot at being the first to discover
Earth-size planets. In an article in *the New York Times*, the Kepler
team leader William Borucki is quoted as justifying the extension
as a way of guarding against false claims of discovery by other
astronomers, saying that "If we say, 'Yes, they are small planets,'
you can be sure they are." In February of 2011 the Kepler team
announced that they had, indeed, discovered five Earth-size planets.

Although practicing science in the open is, on balance, preferable,
Borucki isn't totally wrong to be concerned about false claims. On
July 8, 2010, the particle physicist and blogger Tommaso Dorigo
used his blog to report rumors that the long sought after Higgs
particle had finally been discovered. His post emphasized that he
was just repeating unconfirmed rumors, but despite this caveat the
rumors on his blog were picked up by the mainstream media, and
led to articles in places such as the *Daily Telegraph* (UK) and *New
Scientist* magazine. Just nine days later, on July 17, Dorigo used
his blog to retract the rumor: it was a false alarm. Some scientists
criticized Dorigo, claiming that he acted irresponsibly, or was just
looking for notoriety. But scientific rumors are a staple of scientific
life, the kind of thing that scientists talk about over lunch or in
the hall. Indeed, it's through this kind of speculative discussion that
new ideas are often born. And so it was a natural topic to bring up
in the informal environment of a blog, where Dorigo could talk it
over with his particle physicist friends and colleagues. Given this, it's

tempting to instead criticize the mainstream media for irresponsible reporting. But that's also not fair. Dorigo is a professional physicist, well known and well connected in the particle physics community, someone who could be presumed to be in the know. Of course the mainstream media picked up these rumors.

There's a genuine tension here. Blogs are a powerful way to scale up informal scientific conversation, and to explore speculative ideas. But when this exploration is carried out in the open, there is a danger that the mainstream media, eager for a scoop, will spread news of that speculation, creating the impression that it is fact. Fortunately, this is a problem of limited scope. The mainstream media aren't interested in most scientific discoveries, and for those few discoveries that are of broad interest, events like the Dorigo-Higgs incident will help make the media more cautious about reporting unconfirmed rumors. Although people are often cynical about journalism, most major media organizations are acutely aware of their reputation for credibility (or otherwise), and embarrassed if they have to make frequent public retractions. News of the Dorigo-Higgs retraction was carried by more than half a dozen major media organizations, many of which pointed out that the rumor was originally spread by the *Telegraph* and *New Scientist*. That's not the kind of publicity the *Telegraph* and *New Scientist* want. With all that said, we will see this problem more and more in the future. It seems a relatively small price for the benefits of open science.

Won't increasing the scale of science make it harder to verify scientific discoveries? As open science enables us to scale up the process of discovery, the nature of scientific evidence will change, and become more complex. In the case of some discoveries, understanding the evidence in detail may be beyond the ability of any single person. An early example of this occurred in 1983, when mathematicians announced the solution of an important mathematical problem, known as the classification of the finite simple groups. The proof took nearly 30 years to complete, from 1955 to 1983, and involved 100 mathematicians writing approximately 500 journal articles. Many minor gaps were subsequently found in the proof, and at least one serious gap, which has now been resolved (we hope!) by a two-volume, 1,200-page supplement to the proof. In the 1980s, it was unusual for a scientific discovery to have evidence of such complexity. Today it is becoming common. To pick just one source of complexity, consider that modern experiments in many scientific fields are increasingly likely to use hundreds of thousands or even millions of lines of computer code. It's nearly impossible to eliminate all the bugs from such code. How can we be sure the results output by that code are valid? How can other scientists verify and reproduce the results from such experiments? Furthermore, the situation is getting more challenging, as our computational systems become more complex. Single software programs are increasingly being replaced by software ecologies, complex networks of interacting programs, sometimes maintained by many people across many locations. How can we guarantee that such software ecologies will produce reliable and reproducible results? These and other similar concerns affect discoveries ranging from particle physics to climate science, biology to astronomy. It's a kind of science beyond individual understanding. As this new scale of evidence becomes the norm, our standards of evidence will need to evolve. I'm optimistic, though, that we'll rise to the challenge, using our amplified collective intelligence not only to make new discoveries, but also to develop improved methods for testing and validating those discoveries.

Practical Steps toward Open Science

What practical steps can we take toward open science? Worldwide, our governments spend more than 100 billion dollars each year on basic research. That's *our* money, and we should demand a change to a more open scientific culture. I believe that publicly funded science should be open science. Let's look at some practical steps that everyone, from working scientists to members of the general public, can take toward this end.

What can you do if you're a scientist? Try out open science! Upload some of your old data and old code, online. Document it, encourage other people to use it, and make sure you tell them how you'd like to be cited. Try out blogging. Push your comfort zone—try using your blog to develop some of those ideas you've had in the back of your head for years, but never quite got around to pursuing. You've little to lose, and working in the open may breathe new life into your ideas. If that's too much time commitment, try making a few small contributions to others' open science projects—say, making a comment on a science blog, or a contribution to a wiki. Those contributions may be small, but your scientific colleagues will notice, and it will help legitimize the new tools in the scientific community. And you may find it more rewarding than you think. If you're adventurous, try pushing the boundaries. Ask yourself if you can pioneer a new way of doing science, as the Polymath Project, Foldit, and Galaxy Zoo have done. What can you conjure with imagination and determination? Even if your ventures in open science aren't successful, think of your efforts as service to your community. And, of course, you don't need to do all your science in the open, or even more than a small fraction.

Above all, be generous in giving other scientists credit when they share their scientific knowledge in new ways. Find ways to cite the ideas and data and code they share online. Encourage them to promote their open work, to highlight it on their curriculum vitae and on their grant applications, and to find ways of demonstrating its impact. This is the way to get new citation-measurement-reward cycles going. Of course, you will at times encounter colleagues with old-fashioned scientific values, people who are dismissive of new ways of sharing knowledge, and who think that the only measure of success for scientists is how many papers they've published in high-profile journals such as *Nature*. Talk with those people about the value of new ways of sharing knowledge, and of the courage it takes for scientists, especially young scientists, to work in the open. Sharing ideas and code and data openly, online, is every bit as important as publishing papers, and it is only old-fashioned values that say otherwise.

If you're a scientist who is also a programmer, you have a special role to play, an opportunity to build the new tools that redefine how science is done. Be bold in experimenting with new ideas: this is the golden age of scientific software. But also be bold in asserting the value of your work. Today, your work is likely to be undervalued by old-fashioned colleagues, not because of malice, but because of a lack of understanding. Explain to other scientists how they should cite your work. Work in cahoots with your scientist programmer friends to establish shared norms for citation, and for sharing of code. And then work together to gradually ratchet up the pressure on other scientists to follow those norms. Don't just promote your own work, but also insist more broadly on the value of code as a scientific contribution in its own right, every bit as valuable as more traditional forms.

What if you work at a grant agency? Talk to people in the scientific communities you serve, and ask what knowledge is currently locked up inside scientists' heads and laboratories. What tools would be most effective for sharing that knowledge? Is there an opportunity to develop policies on open access, open data, and open code? How can we go beyond today's open access and open data policies? Can we use examples such as the arXiv and SPIRES as models to help create new norms for citation and new tools for measurement, and so expand science's reputation economy? More generally, if you're involved in government or in the policy-making process, then you can help by getting involved, by lobbying for open access and open data, and more generally by raising awareness of the issue of open science.

And what can you do if you are not a scientist, don't work for a grant agency, and don't work in policy, but are a citizen with an interest in science and human welfare? Talk with your friends and acquaintances who are scientists. Ask them what they're doing to make their data open. Ask them what they do to share their ideas publicly and rapidly. Ask them how they share their code. For open science to succeed, what's needed is a change in the values of the scientific community. If all scientists believe wholeheartedly in the value of working in the open, online, then change will come. This is fundamentally a problem of changing hearts and minds. There is no stronger force for achieving such a change than raising public awareness, so that *everyone* in our society understands the tremendous value of open science, and understands that achieving open science is one of the great challenges of our age. If every scientist in the world is being asked by their friends and family what they're doing to make science more open, then change will come. If every grant agent and every leader at our universities is being asked by *their* friends and family what they're doing to make science more open, then change will come. And if pressure is put on our politicians by a public demanding a more open scientific culture, then change will come.

This needs to become an issue of general concern to our society, a political issue and a social issue that is understood by everyone to be of critical importance. You can help achieve this by using your personal power, your connections, and your imagination to lobby politicians and grant agencies to make policies that encourage openness. (I've listed some organizations already doing this in the "Selected Sources and Suggestions for Further Reading" section at the end of this book.) What types of knowledge will we, as a society, expect and incentivize scientists to share with the world? Will we continue with our current approach? Or will we choose to create a scientific culture that embraces the open sharing of knowledge, the development of new tools that extend our problem-solving ability and speed up scientific discovery?

The steps I have just described are all small steps. But together they will create an irreversible movement toward more open ways of doing science. The inventor and scientist Daniel Hillis has observed that "there are problems that are impossible if you think about them in two-year terms—which everyone does—but they're easy if you think in fifty-year terms." The problem of open science is a problem of this type. Today, creating an open scientific culture seems to require an impossible change in how scientists work. But by taking small steps we can gradually cause a major cultural change.

The Era of Networked Science

I wrote this book with the goal of lighting an almighty fire under the scientific community. We're at a unique moment in history: for the first time we have an open-ended ability to build powerful new tools for thought. *We have an opportunity to change the way knowledge is constructed.* But the scientific community, which ought to be in the vanguard, is instead bringing up the rear, with most scientists clinging to their existing way of working, and failing to support those who seek a better way. As with the first open science revolution, as a society we need to actively avert this tragedy of lost opportunity, by incentivizing and, when appropriate, compelling scientists to contribute in new ways. I believe that with hard work and dedication, we have a good chance of completely revolutionizing science.

When we look back at the second half of the seventeenth century, we can see that one of the great changes of that time was the invention of modern science. When the history of the late twentieth and early twentyfirst centuries is written, we'll see this as *the* time in history when the world's information was transformed from an inert, passive state, and put into a unified system that brings that information alive. The world's information is waking up. And that change gives us the opportunity to restructure the way scientists think and work, and so to extend humanity's problem-solving ability. We are reinventing discovery, and the result will be a new era of networked science that speeds up discovery, not in one small corner of science, but across all of science. That reinvention will deepen our understanding of how the universe works and help us address our most critical human problems.

Appendix: The Problem Solved by the Polymath Project

The Polymath Project aimed to prove a mathematical result known as the density Hales-Jewett (DHJ) theorem. Although the proof of DHJ is complex, the basic statement can be understood by anyone. Take a look at the following three-by-three grid:

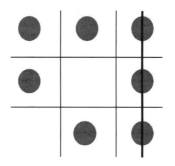

I've marked seven of the squares on the grid with a dot; as you can see, it's possible to draw a line through three of those dots. By contrast, the configuration in the following picture is *line-free*—you can't draw a line through any three of the dots:

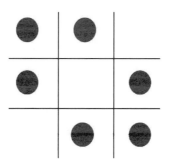

If you play around a bit, you'll discover that this configuration is the largest possible line-free configuration. In particular, if you mark seven dots on the grid, then no matter how you place the dots, it is always possible to draw a line through three of the dots, somewhere on the grid.

Imagine now that we extend the grid to three dimensions, i.e., a three-by-three-by-three grid. It turns out that in three dimensions the largest possible line-free configuration has 16 locations filled in. If we fill in 17 locations on the grid, then no matter which locations we fill in, it will be possible to draw a line through three dots somewhere on the grid. You can take my word for this, or, if you prefer, with a bit of work and three-dimensional imagination, you can convince yourself that this is the case.

Let's make a leap now, and imagine extending the grid from three dimensions to an arbitrary number of spatial dimensions. We'll give the number of dimensions a label—we'll call it n. This extension is hard to visualize, hard enough that most mathematicians can't do it, and they instead translate the problem into an algebraic form. I won't do the algebraic translation here, but instead I'll just explain the question we're concerned with: what's the size of the largest line-free configuration on a grid in n dimensions? We'll give that size a name, calling it s_n. Our discussion above indicates that $s_2 = 6$ and $s_3 = 16$, the sizes of the largest line-free configurations in two and three dimensions. In higher dimensions it rapidly gets extremely difficult to figure out the value of s_n. Mathematicians have worked out the value of s_2 and s_3, as we've seen, and also, with more effort, s_4, s_5, and s_6, but no one in the world knows what the exact value is for s_7. And the situation gets even more complicated in still higher dimensions. But even though it is difficult to figure out an exact value for s_n, the DHJ theorem gives us some partial information about how large s_n is.

In particular, one consequence of the DHJ theorem is that as the number of dimensions n gets very large, the size s_n of the largest line-free configuration is only a tiny fraction of the total number of locations on the grid. Put another way, as n gets large, filling in even a tiny fraction of the grid forces a line somewhere. It doesn't matter how clever you are in filling in locations, there will be a line somewhere. To put the statement in slightly more formal terms, the

DHJ theorem tells us that the fraction of the grid $s_n/3^n$ occupied by the largest line-free configuration gets vanishingly small as n becomes large—it goes to zero in the limit of large n, to use the mathematical lingo.

This is an astonishing statement. As we've seen, in two and three dimensions we can fill in most of the grid before we're forced to put three pieces in a line. Yet in high dimensions, DHJ tells us that a line is forced somewhere on the grid, even if only a tiny fraction of the grid is filled in. It's not at all obvious that this should be the case, and yet the DHJ theorem tells us that it's true.

I've been describing consequences of the DHJ theorem, in order to give you the flavor of what DHJ says. In fact, the full statement of the DHJ theorem is stronger than the consequences I've described so far. It doesn't just work for three-by-three-by ... grids, an analogous statement is true for m-by-m-by ... grids, where m is any number at all. Furthermore, DHJ even tells us that the line will be a certain special type of line called a *combinatorial* line. I won't define combinatorial lines here—see the references in the endnotes if you'd like an explanation of what a combinatorial line is. For now, it's enough that they're a special type of line. What the full statement of the DHJ theorem says is that as the number of dimensions n gets large, the fraction of the $m \times m \times \ldots$ grid occupied by the largest subset without a combinatorial line goes to zero. Put another way, as n gets large, filling in even a tiny fraction of the grid will force a combinatorial line somewhere.

Why should you care about DHJ? If you're coming to DHJ without a lot of mathematical background knowledge, it perhaps seems like an obscure problem. DHJ seems like the kind of puzzle that might make a potentially fun (if difficult) diversion, if you have a puzzle-solving mind. But why is the DHJ theorem any more important than solving a Sudoku puzzle?

Appearances are deceiving. DHJ is a deep theorem. It turns out to have as a consequence many other important and hard-to-prove results of mathematics, some in areas that appear quite unrelated. Think of it as a domino: when it falls, it causes many other important and otherwise hard-to-budge mathematical dominoes to also fall. Let me give you an example of the way DHJ connects to another part of mathematics that seems unrelated—the problem of understanding the structure of the prime numbers. It turns out that DHJ implies a deep result of number theory called Szemerédi's theorem. That theorem was first proved in 1975 by the mathematician Endre Szemerédi; mathematicians have since found several additional proofs. Using ideas drawn from several of those proofs, in 2004 the mathematicians Ben Green and Terence Tao proved a major new result about the structure of the prime numbers. To understand what the Green-Tao theorem says, consider the sequence of numbers 199, 409, 619, 829, 1039, 1249, 1459, 1669, 1879, 2089. These are all prime numbers, and they're evenly spaced; each member of the sequence is 210 larger than the one that precedes it. What the Green-Tao theorem says is that you can find evenly spaced sequences of prime numbers of any length whatsoever. Want an evenly spaced sequence of a million prime numbers? Green-Tao guarantees that such a sequence exists. The theorem doesn't actually give an easily usable recipe for finding such a sequence, but it guarantees that if you search for a sequence long enough, you'll find it eventually. Now, results about the prime numbers probably seem quite unrelated to worrying about line-free configurations in high dimensions. And yet the DHJ-Szemerédi and Szemerédi–Green-Tao connections suggest that there really is a connection between DHJ and the structure of the prime numbers.

The DHJ theorem was first proved in 1991 by the mathematicians Hillel Furstenberg and Yitzhak Katznelson. So when Tim Gowers proposed the Polymath Project, he wasn't proposing that the polymaths find the first proof of DHJ. Rather, he was proposing that they find a new proof. You may be surprised that a top mathematician such as Gowers would be interested in finding a new proof of an already known result. But the existing proof of DHJ used indirect and rather advanced techniques from a branch of mathematics called ergodic theory. While it was a perfectly good proof, Gowers believed that additional insight into DHJ could be gained by finding a new proof that relied on different techniques. In particular, Gowers was interested in finding a proof that relied only on elementary techniques, that is, techniques that didn't require sophisticated mathematics such as the tools of ergodic theory. Sometimes, finding new proofs can give us significant new insights that help us understand why a result is true in the first place. Indeed, this is exactly what happened with the multiple proofs of Szemerédi's theorem. When Green and Tao proved their theorem about prime numbers, they drew on ideas from several different proofs of Szemerédi's theorem. That made finding a new proof of the DHJ theorem using only elementary techniques a challenging and worthwhile goal for the Polymath Project.

Acknowledgments

In writing this book I've benefited enormously from the enthusiasm, insight, and support of many people. Especial thanks to Peter Tallack, my agent, whose enthusiasm for the project, perceptive feedback, and knack for asking the right questions has dramatically improved the book. Many thanks also to the team at Princeton University Press, both for their enthusiasm, and for patiently helping me turn this book into a reality. I'm particularly grateful to my editor, Ingrid Gnerlich, as well as Jodi Beder, Bob Bettendorf, Christopher Chung, Kathleen Cioffi, Peter Dougherty, Jessica Pellien, and Julie Shawvan. Thanks to Simon Capelin, Kelly McNees, and Lee Smolin for helpful comments on early versions of my book proposal. Lee Smolin's words of encouragement were especially welcome at a time when I was close to abandoning the book entirely. A big thank you to Eva Amsen, Rob Dodd, Eric Drexler, John Dupuis, Hassan Masum, Christina Pikas, Dorothea Salo, Lee Smolin, and Rob Spekkens for providing thoughtful comments on a draft of the entire book. Rob Spekkens not only provided detailed comments on the entire book, but made several broad suggestions that dramatically improved the entire work. Thanks to Harvey Brown, Amy Dodd, Danielle Fong, Chris Ing, Chris Lintott, Garrett Lisi, Cameron Neylon, Tobias Osborne, Peter Rohde, Mickey Schafer, Carlos Scheidegger, Arfon Smith, John Stockton, and Mark Tovey for detailed comments on early drafts of one or more chapters. Chapters 8 and 9 of this book are adapted in part from an essay that I wrote for my blog [152], and which was subsequently republished in *Physics World* [150]. My thanks to Joao Medeiros and Matin Durrani of *Physics World* for help with the article, and for making this possible. My work has been greatly enriched by the many people in the online open science community, and my thanks go to everyone in that community. Thanks especially to Cameron Neylon, Peter Suber, and the many others who have done so much to create a thriving online community for open science. Thanks also to those many other people whose insights have informed my thinking, including: Scott Aaronson, Hal Abelson, Richard Akerman, Dave Bacon, Gavin Baker, Travis Beals, Pedro Beltrao, Mic Berman, Michael Bernstein, Peter Binfield, Robin Blume-Kohout, Jean-Claude Bradley, Björn Brembs, Titus Brown, Zacary Brown, Howard Burton, Carl Caves, Ike Chuang, Ken Coates, Alessandro Cosentino, John Cumbers, Wim van Dam, Amy Dodd, Rob Dodd, Michael Duschenes, Drew Endy, Steven van Enk, Steve Flammia, Connie French, Chris Fuchs, Joshua Gans, Alexei Gilchrist, Benjamin Good, Daniel Gottesman, Tim Gowers, Christopher Granade, Ilya Grigorik, Nicholas Gruen, Melissa Hagemann, Timo Hannay, Aram Harrow, Andrew Hessel,

Daniel Holz, Tad Homer-Dixon, Bill Hooker, Sabine Hossenfelder, Jonathan Hunt, Heather Joseph, Jason Kelly, Marius Kempe, Manny Knill, Steve Koch, Matt Leifer, Hope Leman, Daniel Lemire, Debbie Leung, Mike Loukides, Sean McGee, Bob McNees, Hassan Masum, Gerard Milburn, Len Mlodinow, Peter Murray-Rust, Brian Myers, Béla Nagy, Anders Norgaard, Jill O'Neill, Tobias Osborne, Seb Paquet, Heather Piwowar, Jorge Pullin, Srinivasan Ramasubramanian, Neil Saunders, Kevin Schawinski, Cosma Shalizi, Alice Sheppard, John Sidles, Deepak Singh, Rolando Somma, Hilary Spencer, Graham Steel, Victoria Stodden, Dan Stowell, Brian Sullivan, Pawel Szczęsny, Terry Tao, Kaitlin Thaney, Matthew Todd, Ben Toner, Umesh Vazirani, Ricardo Vidal, Christian Weedbrook, Andrew White, John Wilbanks, Greg Wilson, Shirley Wu, Carl Zimmer, and Bora Zivkovic. To those other people whose names should be on this list, but which aren't, my apologies, and my thanks. More than I can say, thankyou to my family, for their love and support: Howard and Wendy Nielsen; Stuart and Shelly Nielsen; Kate Nielsen and Scott Andrews; my niece and nephews Zie, Cooper, Blake, and Bowen; my Grandma Ricardo; Rob and Diane Dodd; Amy Dodd; and Candace and Jonny Marzano. My greatest thanks of all go to my wife, Jen Dodd. Jen's comments and criticisms have greatly improved every aspect of this book, from the tiniest details to the large-scale structure and the overall argument. Without her encouragement and support I would never have begun the book, and certainly would never have finished it.

Selected Sources and Suggestions for Further Reading

This book is in considerable part a work of synthesis, and it owes a tremendous debt to the work of others. Detailed notes on my sources can be found beginning on page 221. Here, I describe a few of the sources that have most decisively influenced my thinking, and suggest further reading.

Collective intelligence: The idea of using computers to amplify individual and collective human intelligence has a long history. Influential early works include Vannevar Bush's celebrated article "As We May Think" [31], which described his imagined memex system, and inspired the seminal work of both Douglas Engelbart [63] and Ted Nelson [145]. Although these works are many decades old, they lay out much of what we see in today's internet, and reveal vistas beyond. Aside from these foundational works, my ideas about collective intelligence have been strongly influenced by economic ideas. Herbert Simon [197] seems to have been the first person to have pointed out the crucial role of attention as a scarce resource in an information-rich world. I also greatly enjoyed Michael Goldhaber's provocative article [75] on "The Attention Economy and the Net." Complementing this is the work of complexity theorist Scott Page demonstrating the value of cognitive diversity in group problem solving [168], and Hayek's notion of "hidden knowledge" and the use of prices as signals to aggregate that knowledge [93]. Other influential works on related subjects include Hutchins's detailed anthropological analysis of collective intelligence in the navigation of a ship [95], Lévy's book on collective intelligence [124], and the stimulating collection of essays on collective intelligence recently assembled by Mark Tovey [224]. Writing from a very different point of view, David Easley and Jon Kleinberg have written a great textbook, *Networks, Crowds, and Markets* [59], which summarizes much of the mathematical and quantitative research on networks. Finally, I recommend Nicholas Carr's book *The Shallows* [35]. It asks the fundamental question, how are online tools changing the way we (individually) think? I believe Carr's answer is incomplete, but it's a stimulating exploration of this important question.

Open source: The best way to get informed about open source is to participate in some open source projects. You can also learn a great deal by reading over the code and discussion archives from open source projects such as Linux and Wikipedia. While writing this book I spent many happy hours doing just that, and can tell you that not only is it informative, it's often surprisingly fun, a kind of cheap entertainment for geeks. I also recommend taking a good look at GitHub (http://github.com), which is the most important current locus for open source work. A good overview of open source is Steven Weber's *The Success of Open Source* [235]. Its only drawback is that it's becoming a little dated (2004), but there is much in the book that is relatively timeless. Going even further back, there is Eric Raymond's famous essay "The Cathedral and the Bazaar" [178]. Raymond's essay is what first got me (and many others) interested in open source, and it remains well worth reading. Yochai Benkler's insightful "Coase's Penguin, or, Linux and *The Nature of the Firm*" [12] and *The Wealth of Networks* [13] have strongly influenced much thinking about open source, especially in the academic community. Finally, I recommend Ned Gulley and Karim Lakhani's fascinating account [87] of the Mathworks programming competition.

Limits to collective intelligence: Informative summaries are Cass Sunstein's *Infotopia* [212] and James Surowiecki's *The Wisdom of Crowds* [214]. Classic texts include Charles Mackay's *Extraordinary Popular Delusions and the Madness of Crowds*, first published in 1841, and since reprinted many times [130], and Irving Lester Janis's *Groupthink* [99]. Of course, a considerable fraction of our written culture deals, directly or indirectly, with the challenges of group problem solving. Among the more formative accounts for me were Ben Rich's *Skunk Works* [184], Richard Rhodes's *The Making of the Atomic Bomb* [183], and Robert Colwell's *The Pentium Chronicles* [45]. A little further afield, Peter Block's book *Community: The structure of belonging* [18] contains many insights about the problems of building community. And, finally, Jane Jacobs's masterpiece *The Death and Life of Great American Cities* [98] is a superb account of how very large groups tackle a core human problem: how to make a place to live.

Networked science, in general: The potential of computers and the network to change the way science is done has been discussed by many people, and over a long period of time. Such discussion can be found in many of the works described above, in particular the work of Vannevar Bush [31] and Douglas Engelbart [63]. Other notable works include those of Eric Drexler [57], Jon Udell [227], Christine Borgman [23], and Jim Gray [83]. See also Tim Berners-Lee's original proposal for the world wide web, reprinted in [14]. A stimulating and enjoyable fictional depiction of networked science is Vernor Vinge's *Rainbows End* [231].

Data-driven science: One of the first people to understand and clearly articulate the value of data-driven science was Jim Gray, of Microsoft Research. Many of his ideas are summarized in the essay [83], which I also mentioned above. That essay is part of a stimulating book of essays entitled *The Fourth Paradigm* [94]. The book is freely downloadable from the web, and gives a good overview of many parts of data-driven science. Another thought-provoking article is "The Unreasonable Effectiveness of Data" [88], by Alon Halevy, Peter Norvig, and Fernando Pereira. All three of the authors work for Google, which has perhaps the most data-driven culture of any organization in the world, and the article conveys well the radical shift in perspective that comes from thinking in a data-driven way. If you have a background in programming, I also recommend Norvig's terrific short essay [157] on how to write a (data-driven, naturally!) spelling corrector in just 21 lines of code. There are many, many texts and papers on topics related to data-driven intelligence.

(Note, though, that most don't use the term.) A good practical introduction is Toby Segaran's *Programming Collective Intelligence* [191].

The democratization of science and citizen science: The democratization of science has analogs in the business world, in phenomena such as user-generated innovation, and open innovation models for business. See, for example, Eric von Hippel's book *Democratizing Innovation* [233], whose title inspired the title of chapter 7, and Henry Chesbrough's *Open Innovation* [36]. The point of view developed in chapter 7 also owes a great deal to Clay Shirky's notion that our society has a cognitive surplus [195, 194; see also 196] which can be used in the service of new forms of collective action.

Open science: My analysis of open science is strongly influenced by the work of Mancur Olson [161] on collective action, and by the work of Elinor Ostrom [165] on the management of common pool resources such as fisheries and forests. Both these works have many more implications for open science than I have described. In particular, I only briefly touched on many of the detailed principles that Ostrom identifies for the management of common pool resources. Many of those principles can be fruitfully applied or adapted to open science. I have also been stimulated by the work of Robert Axelrod [9] on the conditions under which parties will cooperate; the problem of large-scale cooperation is an example of a collective action problem. On the early history of open science, I've been stimulated by many sources, but especially by Paul David [49], Elizabeth Eisenstein [61], and Mary Boas Hall [89].

One thing that pained me while writing this book is that narrative constraints meant that I've had to omit nearly all the *thousands* of open science projects now going on. Fortunately, there are many excellent sources for keeping track of what's going on in open science today. Let me mention just a few. One of the most valuable is Peter Suber's website (http://www.earlham.edu/~peters/hometoc.htm), which is a tremendous resource on all aspects of open science, but especially open access publishing. Suber's superb blog (http://www.earlham.edu/~peters/fos/fosblog.html) is no longer updated, but remains a valuable historical resource. And Suber's ongoing Open Access Newsletter (http://www.earlham.edu/~peters/fos/newsletter/archive.htm) is essential. Another excellent source on open science is the blog of Cameron Neylon (http://cameronneylon.net/). Neylon is one of the pioneers of open notebook science, and has many stimulating things to say about open science more generally. You can also find many open scientists and open science projects using services such as Twitter and FriendFeed. A good entry into this world is to use Google to search for "Twitter open science."

In addition to these individuals, there are many organizations working for open science. The Alliance for Taxpayer Access (http://www.taxpayeraccess.org/) has lobbied the US government for policies on open access to scientific papers and scientific data. For instance, it was in part through their lobbying that the NIH open access policy described in chapter 7 came about. Other organizations working toward open science include Science Commons (http://sciencecommon.sorg), which is part of the Creative Commons organization, and the Open Knowledge Foundation (http://okfn.org).

The challenge of creating a more open culture is not limited to science. It's also being confronted in general culture. People such as Richard Stallman [202], Lawrence Lessig [122], and many others have described the benefits openness brings in a networked world. They've developed tools such as Creative Commons licensing (http://creativecommons.org) and "copyleft" licenses to help bring about a more open culture. My thinking has been especially strongly influenced by Lessig [122]. However, although open science has many parallels to the open culture movement, science faces a unique set of forces that inhibit open sharing. That means that tools such as Creative Commons licenses, which have been tremendously effective in moving to a more open culture, don't directly address the principal underlying challenge in science: the fact that scientists are rewarded for publishing papers, and not for other ways of sharing knowledge. So although open science can learn a lot from the open culture movement, it also requires new thinking.

Notes

Some of the references that follow include webpages whose URLs may expire after this book is published. Such webpages should be recoverable using the Internet Archive's Wayback Machine (http://www.archive.org/web/web.php). Online sources are often written informally, and I've reproduced spelling and other errors verbatim when quoting such sources.

Chapter 1. Reinventing Discovery

p 1: Gowers proposed the Polymath Project in a posting to his blog [79]. For more on the Polymath Project, see [82].

p 2: Gowers's announcement of the probable success of the first Polymath Project: [81].

p 2 The Polymath process was "to normal research as driving is to pushing a car": [78].

p 3: The term *collective intelligence* was introduced by the philosopher Pierre Lévy [124]. A stimulating recent attempt to measure collective intelligence and to relate it to qualities of participants in the group is [243].

p 3 the process of science will... change more in the next twenty years than it has in the past 300 years: the author Kevin Kelly has made a similar claim in [108] (see also [109]): "There will be more change in the next 50 years of science than in the last 400 years." There is some broad overlap in my reasoning and Kelly's, e.g., we both emphasize the importance of collaboration and large-scale data collection. There are also some considerable differences in our reasoning, e.g., Kelly emphasizes changes such as triple-blind experiments, and more prizes in science, while I believe these will play a comparatively minor role in change, and that the following three areas are the most critical: (1) collective intelligence and data-driven science, and the way they change how science is done; (2) the changing relationship between science and society; and (3) the challenge of achieving a much more open scientific culture.

p 4: GenBank is at http://www.ncbi.nlm.nih.gov/genbank/. The human genome is available at http://www.ncbi.nlm.nih.gov/projects/genome/assembly/grc/human/index.shtml, and the haplotype map is available at http://hapmap.ncbi.nlm.nih.gov/.

p 7: A firsthand account of the Bermuda meeting, including a statement of the Bermuda Agreement, may be found in [211]. The Clinton-Blair statement on sharing of genetic data doesn't explicitly name the Bermuda Agreement, but the principles espoused are essentially the principles agreed on in Bermuda. The statement may be found at [102].

p 7: I've used the Bermuda Agreement as an example of a collective agreement that drives data sharing. In fact, the amount of genetic data deposited in GenBank has doubled roughly once every 18 months since GenBank was founded, and this trend was not noticeably hastened by the Bermuda Agreement. You might wonder if the Bermuda Agreement was truly all that important to increased data sharing. Of course, part of the increase in data sharing is due to better sequencing technology. But the increase is also due in part to a broad drive by the biological community to share data more freely. The Bermuda Agreement is merely part of that broad drive, albeit perhaps the most visible manifestation.

p 7: On extensions of the Bermuda Agreement, see especially the Fort Lauderdale Agreement [237].

p 7: On the sharing of influenza data, see for example [20] and [60] on the avian flu outbreak of 2006, and [32] on the swine flu pandemic of 2009–10.

p 10 We are living in the time of transition to the second era of science: A related claim has been made by the database researcher Jim Gray [83] (see also the volume in which Gray's essay appears [94]). Gray has claimed that we are today entering what he calls a "fourth paradigm" of scientific discovery, one based around highly data-intensive science in which computers help us find meaning in data. In Gray's account this fourth paradigm is an extension of what he calls the first paradigm (empirical observation), second paradigm (the formation of models to explain observation), and third paradigm (the use of simulation to understand complex phenomena) of science. It's true that data-intensive science is important, and we'll discuss it in chapter 6. But Gray's conception of the current change in science is too narrow. Science is about much more than just finding meaning in data. It's also about the ways in which scientists work together to construct knowledge, and how the scientific community relates to society as a whole. Those aspects of science are also being transformed by online tools. Furthermore, each of these shifts impacts on and reinforces the others. So, for example, to really understand the impact of data-intensive science we must understand changes in the ways scientists work together. Gray's fourth paradigm is just part of the changes being wrought by networked science.

Chapter 2. Online Tools Make Us Smarter

p 15: My account of Kasparov versus the World is based primarily on Kasparov's book (with Daniel King) [107], and Irina Krush's account of the game (with Kenneth Regan) [115].

p 15 "the greatest game in the history of chess": from a Reuters interview with Kasparov conducted during the game [186], at move number 37. It is part of an interesting longer comment by Kasparov: "'It is the greatest game in the history of chess. The sheer number of ideas, the complexity, and the contribution it has made to chess make it the most important game ever played."

p 19: James Surowiecki, *The Wisdom of Crowds*, [214].

p 20: Nicholas Carr's book *The Shallows* [35] is an expanded version of an earlier article, "Is Google Making Us Stupid?" [34]. Related arguments have also been made by Jaron Lanier [117].

Chapter 3. Restructuring Expert Attention

p 22: On ASSET India, InnoCentive, and Zacary Brown: [29, 222]. The text on InnoCentive is a much expanded and adapted version of text from my article [153].

p 23 Many of the successful solvers report, as Zacary Brown did, that the Challenges they solve closely match their skills and interests: see [116] for more on the characteristics of successful solvers. Note that this study also found that people often solve Challenges that are nominally outside their domain of expertise. A chemist might, for example, solve a problem in biology. This seems like a contradiction to the claim about a close match to expertise, but it is not: the key difficulty in solving the biological problem might be a very specific piece of expertise from chemistry. So when one looks at the Challenge solutions at a fine-grained level, the match to expertise is often exceptionally close.

p 24 It's because Zacary Brown has such an enormous comparative advantage that he and ASSET can work together for mutual benefit: "comparative advantage" is a technical term from economics, and I'm using the term in that sense. Elsewhere, when I speak of people applying their expertise in the "best" possible way (or similar language), I mean best in the sense of maximizing comparative advantage, not maximizing absolute advantage.

p 24: The critical character of human attention as a scarce resource in an information-rich world was pointed out in a prescient article by Herbert Simon [197]. A striking speculative work on the economics of attention is the article by Michael Goldhaber [75]. See also [151].

p 27: Regarding the term "designed serendipity," Jon Udell used the term "manufactured serendipity" to describe a similar concept in [228]. I've used "designed serendipity" instead because it emphasizes the way serendipity can be achieved as the result of deliberate design choices. The idea of designed serendipity seems to have originated in the open source software movement, and was succinctly captured in Eric Raymond's [178] observation that when debugging open source software, "given enough eyeballs, all bugs are shallow." Raymond dubbed this observation Linus's Law, after the creator of Linux, Linus Torvalds. We can generalize Linus's Law to other forms of problem solving: "Given enough eyeballs, all problems are easy." It's not literally true, but it does capture something of the essence of designed serendipity.

p 27 "Grossmann, you must help me or else I'll go crazy!": the Einstein-Grossmann story is told in full in [169].

p 30: The discussion of conversational critical mass is inspired in part by chapter 3 of [189].

p 30 Polymath participants often "found [themselves] having thoughts that [they] would not have had without some chance remark of another contributor": [80].

p 31: On the value of cognitive diversity, see, for example, the work of Scott Page [168] and Friedrich von Hayek [93].

p 32: The phrase "architecture of attention" is inspired by Tim O'Reilly's elegant phrase "architecture of participation" [162]. O'Reilly uses his term "to describe the nature of systems that are designed for user contribution." We're interested in systems designed for creative problem solving, and in such systems it is the allocation of expert attention that is most crucial.

p 34: The number of employees on *Avatar* is from [65].

p 36: The 1983 discovery of the Z boson is described in [4].

p 37 "who is in charge of the supply of bread to the population of London?": see Paul Seabright's *The Company of Strangers* [190].

p 37 What makes prices useful is that . . . they aggregate an enormous amount of hidden knowledge: [93].

p 38: The "dumb question" was posed by Polymath participant Ryan O'Donnell: [159].

p 39: On the point that online tools are subsuming and extending both conventional markets and conventional organizations: a related point has been made by the theorist Yochai Benkler in his article "Coase's Penguin, or, Linux and *The Nature of the Firm* [12]." Benkler has a different focus, being concerned not so much with the solution of creative problems as with the production of goods. He proposes that online collaboration has enabled a third form of production, beyond markets and conventional organizations, which he calls "peer production." I believe this is too narrow a point of view, both for creative problem solving and for the production of goods. Online tools can be used to *subsume* both markets and conventional organizations as special cases, and also enable many new forms of production and creative problem solving. Thus it's not that we now have a third form of production. It's that we now have a means of production that includes all our former forms as special cases, and enables new forms.

Chapter 4. Patterns of Online Collaboration

p 44: Insightful accounts of open source software development include [12, 13, 178, 235]. Even more useful are the innumerable open source projects maintained online at sites such as GitHub (http://github.com) and SourceForge (http://sourceforge.net).

p 44: My history of Linux is based largely on postings to the comp.os.minix, alt.os.linux, and comp.os.linux newsgroups in 1991 and 1992. I found reading through those forums surprisingly enjoyable, and even compelling: as you read, you start to get a visceral sense of what was involved in producing a marvel of modern software. My account of Linux was also broadly influenced by [235], as well as many other sources for details (see below).

p 45 Shortly after Torvalds's post...: comp.os.minix newsgroup posting, January 13, 1992.

p 45 80 people were named as contributors in the Linux Credits file: See [226] for the history of the Credits file. March 1994 is the first time such a file was included in Linux.

p 45 By early 2008, the Linux kernel...: [114].

p 45: On the role of Linux in Hollywood animation and visual effects companies, see [90] for an account as of 2002, the time when Linux was coming into the industry, and beginning to dominate. [187] claims that as of 2008, Linux was used on "more than 95% of the servers and desktops at large animation and visual effects companies."

p 45 Open source software projects have two key attributes: Some open source advocates prefer a more nuanced description of open source than the description I've given. Many complex and sometimes heated discussions have gone on regarding which projects should be regarded as truly open source. Indeed, a not-for-profit organization named the Open Source Initiative exists in part to decide whether a project should be labeled open source, and if so, to provide certification. From the outside this may look like pedantic nitpicking, but there are good reasons for it. Open source is sometimes seen as a threat to some large software companies: for instance, Linus Torvalds once said in the *New York Times*, "I'm not out to destroy Microsoft. That will just be a completely unintentional side effect" [52]. Some of the companies threatened by open source have struck back by trying to break the open source brand, releasing products they call "open source," but lacking crucial features found in truly open source projects. In May 2001, Microsoft senior vice president Craig Mundie [142] announced that Microsoft would be releasing some products as "Shared Source," stating that "Shared Source is Open Source." A close look at the Microsoft Shared Source licenses shows that they are heavily skewed toward users of Microsoft products, and in some cases prevent programmers from modifying code. This is certainly not open source! That type of incident shows why open source advocates sometimes get upset when people use the term "open source" in a sloppy way. We'll take a more relaxed approach that I believe gets at the essence of open source, but without getting bogged down in the complexities of whether the projects we describe would pass all the stringent tests demanded by some open source advocates.

p 46: The figure of 4,300 lines of code added to the Linux kernel per day is from an informative talk about the Linux kernel development process, by Greg Kroah-Hartman [113].

p 46 an experienced developer will typically write a few thousand lines of code per year: this estimate is based on the COCOMO II software model [19].

p 46 SourceForge is home to more than 230,000 *open source* projects: [239].

p 46 open source is a general design methodology that can be applied to any project involving digital information: The open source methodology can also be applied to nondigital information. You could imagine, for instance, using architects' printed plans as the basis for open source design of buildings. The problem with analog information is that it tends to degrade as it is repeatedly copied, which limits its usefulness for the open source methodology.

p 46 Open Architecture Network: http://www.openarchitecturenetwork.org. The Open Architecture Network was introduced in a talk by Cameron Sinclair: [198].

p 48: On open source biology, see, for example, chapter 13 of [33].

p 49: My account of the near fork of Linux is based primarily on the online Linux kernel mailing list, with some additional information from [235].

p 51: On the difficulty of making open source development modular, a comment I've sometimes heard from non-programmers who are interested in open source is that programming is "naturally modular." This is a misconception, and seems to be based on a confusion in terminology. It's true that many programming languages encourage a modular structure in development, and for *small* programs this makes modular design easy. But for large-scale systems such as Linux modularity means something quite different, and is much more difficult to achieve. Large-scale systems are no more naturally modular than a painting is naturally modular because paint happens to be built from modular units (molecules). Rather, modularity in large-scale software engineering requires clever design through several levels of abstraction, and that, in turns, requires a strong commitment to the principle of modularity on the part of developers.

p 52: For Linus Torvalds on modularity, see [223].

p 53: The Million Penguins blog is at http://www.amillionpenguins.com/blog/, and has links to other resources associated with the Million Penguins project, including the wiki used to write the novel. I learned of the project from [139], which ran the same excerpt from the novel I have used.

p 55: Firefox's online issue tracker may be found at http://bugzilla.mozilla.org.

p 55: The favicon bug in Firefox is described at https://bugzilla.mozilla.org/show_bug.cgi?id=411966.

p 56 The issue tracker isn't just for fixing bugs, it's also used to propose and implement new features: In fact, the issue tracker is just one of several ways in which Firefox developers can propose new features. Other forums used to propose new features include an online mailing list, a wiki, and even a weekly conference phone call of Firefox developers.

p 58 more than a billion lines: This and the estimate of the rate of code growth are conservative estimates, based on work by Deshpande and Riehle [51], current as of the end of 2006.

p 58: Alan Kay's story about Donald Knuth is from page 101 of [192].

p 59 "Good programmers code; great programmers reuse other people's code": Variants of this saying have floated around the open source world for years, but I haven't been able to track down the original source. This is fitting. There's more, too: the quote is a paraphrase of a quote often attributed to Picasso, "Good artists copy; great artists steal." I haven't been able to find a verifiable source for the Picasso quote, but compare T. S. Eliot's "Immature poets imitate; mature poets steal" [62].

*p 59 For more on the MathWorks competition, see [87] and, especially, [88].

p 61 "I started to become 'obsessed' ": [86].

p 63: A study by two scientists at the software company SAP, Oliver Arafat and Dirk Riehle...: [3].

p 63: You might conclude from the discussion of microcontribution that open source software is mostly built up out of tiny contributions. But just because small contributions are more frequent doesn't mean that they make up the bulk of the final product. It might be that a few large contributions overwhelm the many smaller contributions. And, indeed, in many open source projects that's what happens: tiny changes are the most frequent, but the final product is still dominated by relatively large chunks of code. It's tempting, then, to reverse direction, and conclude that small contributions aren't that important, that really they're a distraction. But that's wrong too, a bit like arguing that *Hamlet* would be a better play with everything removed except the great soliloquies. Both the large and the small contributions are crucial. The large contributions matter for the obvious reason, and the small contributions matter because they move the conversation forward, and help the collaboration explore a broader range of ideas. It's from the best of those ideas that the big contributions spring.

p 66 a collaboration needs to know what the collaboration knows: This observation, often in different guises, seems to have been made many times. I first fully appreciated it after reading [28].

p 67 "If anything in my life that I've participated in...": This quote is from a comment made by commenter AdmiralBumblebee [30] on the website reddit. It's worth mentioning that the comment was stimulated by an early version of the material that opens chapter 2 of this book, which AdmiralBumblebee felt reflected "commercial hype" and a sponsor's view of the game. My account is, however, not based on information from the sponsor, Microsoft, but primarily on the firsthand accounts of Kasparov and Krush, and corroborated by several other sources.

Chapter 5. The Limits and the Potential of Collective Intelligence

p 69: The Stasser-Titus experiments are described in [204], which contains many more details than my abbreviated account. A review of work following up these experiments is [203]. An informative broader summary of the way collective intelligence can fail is Sunstein's book *Infotopia* [212].

p 75 the stronger players on the World Team could usually agree on which analyses were best: There was a significant exception to this, which is that early in the game Microsoft asked the World Team advisors not to consult with one another, and so they did not have the opportunity to come to agreement. But many of the stronger World Team players were in close contact, and they were frequently able to come to agreement.

p 78: On the limits to collective intelligence, and problems such as groupthink, information cascades, etc., see [99, 212, 213, 214], and references therein.

p 79: Regarding the rapid acceptance of Einstein's ideas, it helped that leading scientists such as Lorentz and Poincaré arrived at similar conclusions at about the same time. But although Einstein's formulation of relativity was even more radical than the formulations of Lorentz and Poincaré, it quickly became accepted as the correct way to think about relativity.

p 79: On the discovery of DNA, and Pauling's error, see Watson's memoir, *The Double Helix* [234].

p 80 "If Feynman says it three times, it's right": [72].

p 84: My thanks to Mark Tovey for help constructing this example on optical illusions and cognitive science.

p 85: On collaboration markets, see also [246] and [146].

p 85: The discussion of topological quantum computers is inspired by [22]. Topological quantum computers were originally proposed in a remarkable article by Kitaev [111].

Chapter 6. All the World's Knowledge

p 91: Swanson's discovery of the magnesium-migraine connection is described in [215], and reviewed in [216].

p 92: An interesting question about the migraine-magnesium connection is why it wasn't discovered by, say, scientists working on epilepsy, some of whom were presumably aware of the connection of epilepsy to both migraines and magnesium deficiency. Speculating, it seems likely that the reason this connection went unnoticed is that (1) those scientists were focused mostly on understanding epilepsy, not other conditions; and (2) a single connection linking migraines and magnesium deficiency isn't enough of a pattern to infer anything. Epilepsy is connected to many different conditions, most of which have no direct relation to one another.

p 92: On Swanson's procedure, there is, of course, nothing new about inferring undiscovered knowledge from existing scientific knowledge. It's standard practice in fields like my own field of theoretical physics. But Swanson's systematic computer-mediated application of this idea in medicine was new, and foreshadowed an explosion in the use of similar data mining techniques in many areas of science.

p 93: The notion of the extended mind has been discussed in [43].

p 93: The paper describing the use of Google search queries to track the flu is [71].

p 93: Influenza annual mortality rates are from the World Health Organization [244].

p 93: The Spanish flu mortality rate is from [219].

p 94: The Google Flu Trends website is http://www.google.org/flutrends.

p 94: The CDC/General Electric system for tracking influenza is described in [136].

p 94: The follow-up study showing that Google Flu Trends is better at tracking influenza-like illnesses than it is at tracking laboratory-confirmed cases of influenza is [163].

p 94: On the use of search queries to predict unemployment, see [6]. On the use of search queries to predict housing prices, see [245]. On the use of search queries to help improve predictions for how well songs will do on the charts, see [73]. For a broad range of applications, see [42]. A study using Twitter to predict movie box-office revenue is [7]. Finally, see [11] for a thought-provoking discussion of Google as a "database of [human] intention."

p 95: For Eric Schmidt on privacy, see [64].

p 96: The phrase "unknown knowns" was suggested in this context by Jen Dodd and Hassan Masum, inspired by former US Secretary of Defense Donald Rumsfeld's famous use [188] of "unknown unknowns."

p 97: The discovery of the Sloan Great Wall of galaxies is described in [77]. The Sloan Great Wall galaxies don't appear to be gravitationally bound together, and so some astrophysicists don't regard them as a single structure. However, much of the story told in this section carries over to several other large-scale features of the universe—my choice of the Sloan Great Wall was somewhat arbitrary.

p 100: The discovery of the many dwarf galaxies near to the Milky Way was described in multiple papers. For an overview, see http://www.sdss.org/signature.html.

p 100: The discovery of the orbiting black holes was described in [25]. In the text I state that Boroson and Lauer searched through galaxy images from the SDSS. To be a bit more precise, they searched through a selection of 17,500 quasars, a special type of galaxy known to contain supermassive black holes. For more on what quasars are and why they're interesting, see the description on page 130. Note that there has been considerable follow-up discussion in the astronomy and astrophysics community of whether the discovery in [25] is, in fact, of a pair of orbiting black holes, or perhaps something else. This conversation is ongoing.

p 101: The Sloan Digital Sky Survey was described in [247]. The citation numbers for this paper are from the service Google Scholar. The numbers are conservative, since they do not include citations to subsequent data releases, and many other key papers from the SDSS.

p 102: The SDSS has codified many of their policies about collaboration and data sharing at http://www.sdss.org/collaboration/. It makes surprisingly stimulating reading.

p 102: The SDSS SkyServer is at http://skyserver.sdss.org.

p 104: On Watson, Crick and Franklin, see Watson's memoir, *The Double Helix* [234].

p 105: The webpage for stage III of the SDSS is at http://www.sdss3.org.

p 105: My account of the Ocean Observatories Initiative is based on the project website, at http://www.oceanleadership.org/programs-and-partnerships/ocean-observing/ooi/, and [50].

p 106: Mapping the brain is far too large a subject for me to give a comprehensive list of references. An overview of work on the Allen Brain Atlas may be found in Jonah Lehrer's excellent article [120]. Most of the facts I relate are from that article. The paper announcing the atlas of gene expression in the mouse brain is [121]. Overviews of some of the progress and challenges in mapping the human connectome may be found in [119] and [125].

p 108: Bioinformatics and cheminformatics are now well-established fields, with a significant literature, and I won't attempt to single out any particular reference for special mention. Astroinformatics has emerged more recently. See especially [24] for a manifesto on the need for astroinformatics.

p 113: A report on the 2005 Playchess.com freestyle chess tournament may be found at [37], with follow-up commentary on the winners at [39]. Garry Kasparov's comments on the result are in the fascinating article [106], which contains much of interest on the subject of computers and chess. Additional commentary on Hydra's involvement may be found at [38]. Interestingly, Hydra has played and lost twice in games of correspondence chess, against correspondence chess grandmaster Arno Nickel. Nickel was, however, allowed to use computer chess programs in these games. A full record of Hydra's games may be found at [40].

p 119: Chuck Hansen's book is [92]. The story I recount about Hansen's methodology is told in Richard Rhodes's book *How to Write*, [182], page 61.

p 120: On the semantic web, see [16, 15] and http://www.w3.org/standards/semanticweb/. A stimulating alternate point of view is [88].

p 120: For Obama's memorandum on transparency and open government, see [158].

p 123: The beautiful summary of Einstein's general theory of relativity, "Space-time tells matter how to move; matter tells spacetime how to curve," is due to John Wheeler [240].

p 125 these models have no understanding of the meaning of "hola" or "hello": I use the term "understanding" here in its everyday sense. I suspect, though, that one day we'll discover that what we mean by "understanding" is captured in part (but only in part) by the kind of statistical association in these models.

p 125 no one on the Google Translate team spoke Chinese or Arabic: [69].

p 128: Planck's comment "I really did not give it [the quantum theory] much thought" is from Helge Kragh's article [112].

Chapter 7. Democratizing Science

p 129: My account of Galaxy Zoo is based on the Galaxy Zoo blog, http://blogs.zooniverse.org/galaxyzoo/, the Galaxy Zoo forum, http://www.galaxyzooforum.org, and an article by Chris Lintott and Kate Land [127]. The material on Hanny's Voorwerp draws also on Hanny van Arkel's blog http://www.hannysvoorwerp.com/, and the original discussion thread started by Hanny van Arkel [67]. The first Galaxy Zoo paper on the voorwerp is [128].

p 131: The alternative explanation of the voorwerp is given in [105, 177]. Some comments on the alternative explanation by Galaxy Zoo cofounder and Zookeeper Chris Lintott may be found at [126].

p 135: Alice Sheppard's account of the discovery of the green pea galaxies is in [193]. Note that the galaxy images seen by the Zooites are in false color, and the "green peas" are actually closer to red.

p 138: An enjoyable short article on the discovery of helium is [118].

p 141: Bob Nichol's quote, "I can ask the question 'how many galaxies have a bar through the middle of them' and typically I would embark on a career-long quest to answer this fundamental question . . . ," is from [149].

p 143: Foldit is at http://fold.it. Good overviews of Foldit are [46, 21].

p 147: For Aotearoa on Foldit, see [1] and [2].

p 148: The Foldit results for the 2008 CASP are at [174].

p 149: On John Caister Bennett's discovery of the great comet of 1968, see [104].

p 149: Comet hunter Rainer Kracht's homepage, at http://www.rkracht.de/, has a list of comets he has discovered. Background on SOHO's success at hunting comets may be found at http://sungrazer.nrl.navy.mil/.

p 150: The eBird website is at http://ebird.org, and the project is described in [210]. The information on the number of contributions and contributors is from http://www.avianknowledge.net/content/datasets and [209].

p 150: The open dinosaur project is at http://opendino.wordpress.com/. An overview of the project can be found in [220].

p 151: The use of Galaxy Zoo data to train a computer algorithm is described in [10].

p 153: Clay Shirky's analysis of Wikipedia appeared in [195]. That article is also the origin of the phrase "cognitive surplus." Shirky has developed these ideas at book length in [194].

p 153 On average Americans watch five hours of television per day: [156].

p 154: Clay Shirky's idea of doing "big things for love" is developed at length in his insightful book *Here Comes Everybody* [196]. The quote "We are used to a world where little things happen for love and big things happen for money. . ." is from page 104 of that book.

p 155 "my life changed forever . . . ": [132].

p 155: The 1988 data on polio incidence are from [141].

p 155: Data on 2003 polio incidence are from the Global Polio Eradication Initiative's 2003 annual report, available at http://polioeradication.org.

p 155: The Nigerian boycott of the polio vaccination program is described in [101].

p 156: A review of the literature on the connection between vaccines and autism is [68]. The evidence in this review strongly suggests that there is no causal link.

p 156: The numbers on vaccination rates for measles-mumps-rubella and the rate of measles infection are from [135], based on data from the Health Protection Agency.

p 160: The single best resource on open access is Peter Suber's remarkable blog, Open Access News, available at http://www.earlham.edu/~peters/fos/fosblog.html. The blog was discontinued as of April 2010, but it is well worth browsing through the archives. Suber has prepared an overview of Open Access [207], and a time-line [208], both of which are very helpful for getting a big picture view of open access. Suber and others continue with the Open Access Tracking Project, whose archives may be found at http://oatp.tumblr.com/. For a book-length overview of open access, see [241].

p 161: The arXiv is online at http://www.arxiv.org. Note that the arXiv started in the field of physics, but has since spread to other disciplines, such as mathematics and computer science. In this book I've concentrated on the physics aspects and sometimes refer to it as the physics arXiv, since physics is the field in which the arXiv is most dominant.

p 162: The Public Library of Science (PLoS) website is at http://plos.org. PLoS wasn't the first open access journal, but it was one of the earliest, and I've focused on it because it has blazed trails in many ways.

p 162: For an overview of the NIH Public Access Policy, see [206]. It's short, but contains many informative links.

p 162: The NIH budgetary information is from http://www.nih.gov/about/budget.htm.

p 164: The Elsevier revenue and profit figures are based on the 2009 Reed Elsevier Annual Report [181].

p 164: The American Chemical Society's revenue and profit figures are from [131].

p 164: My account of Eric Dezenhall and the publishers' trade association (the Association of American Publishers) is based on [70], with additional background from [100]. The quotes from PRISM are from [176].

p 165: Simon Singh's original article in which he criticized the British Chiropractic Association (BCA) is [199]. The article by Dougans and Green on the Singh case is at [56]. My discussion also benefited from articles by Ben Goldacre [74] and Martin Robbins [185]. The BCA's description of evidence for the effectiveness of chiropractic treatments is [221]. A similar instance of wiki litigation in the open source software world involved assertions by a company called SCO that code it owned had been incorporated into Linux, as a result of which SCO sued companies such as Novell and IBM. The cases were covered in remarkable detail at a community website called Groklaw (http://groklaw.net), started by a paralegal named Pamela Jones.

p 167: Pharyngula is at http://scienceblogs.com/pharyngula/. The figures for the circulation of the *Des Moines Register* and the *Salt Lake Tribune* are from the Audit Bureau of Circulations [8].

p 170: My account of Easter Island is based on Jared Diamond's book *Collapse* [53]. The reconstruction of Easter Island's history is difficult and complex, and the subject of much contention among scholars; unsurprisingly, some disagree with Diamond's account.

p 171: On the reduction of life expectancy due to HIV/AIDS in the most highly affected African countries, see [103].

p 171: On bridging the ingenuity gap, see [133].

Chapter 8. The Challenge of Doing Science in the Open

p 173: My account of Galileo's work is based upon [238].

p 174: For more on the affair of Galileo and Baldassare Capra, see [17].

p 175: My account of the origins of open sharing of discoveries in science is based in part on Paul David's article [49]. David points out that there is nothing logically inevitable about the emergence of openness in science, and that it was in large part a result of external forces acting on the scientific community, not merely forces within science. David's analysis focuses on the earliest parts of modern science, and emphasizes how prestige seeking by monarchs and other patrons was a motivation for open disclosure of results. In my account I've also emphasized the motivation coming from the public benefit derived from open science. This motivation seems to have acquired more force in later times, as the power of the monarchs diminished.

p 176: The qwiki is online at http://qwiki.stanford.edu/wiki/Main_Page. In my description of the qwiki, I state that only a few pages are regularly updated. In fact, there is a part of the site that receives fairly regular attention: the "Complexity Zoo," a resource for computer scientists that describes different types of computational problem. The Complexity Zoo needs separate consideration, however, for it is based on a project that was originally totally unconnected to the qwiki, and that later merged with the qwiki. As a result, for the purposes of this discussion, I'm treating the "Complexity Zoo" as a separate entity. It is, of course, interesting to ask why the Complexity Zoo succeeded when the rest of the qwiki failed. A full answer to this question is complex, but in brief, the Complexity Zoo has a much narrower scope than the qwiki, and because of this narrower scope a single dedicated person (Scott Aaronson, now of MIT) was able to build it out to the point where it became an extremely useful and well-known resource in the computer science community. The combination of its already high profile and its narrow scope has helped attract a few people to make occasional contributions to its upkeep.

p 176: The term "wiki-science" seems to have been introduced in an essay by Kevin Kelly [108]. Similar ideas were proposed independently (and, in some cases, earlier) by many people. An interesting discussion involving some early contributors to wikis may be found at the Meatball wiki: [137] and [138].

p 178: The job and graduation data for physics are based on the American Institute of Physics' "Latest Employment Data for Physicists and Related Scientists," available at http://www.aip.org/statistics/. I picked physics because reliable data are available. Anecdotal impressions from other fields confirm that the situation is similar.

p 178 Those science wikis that do succeed are usually in a supporting role for some more conventional project: a notable exception to this rule is the Gene Wiki, a successful wiki-based project to annotate genes. Part of what has helped the Gene Wiki succeed is that it is not an independent wiki, but rather a subproject of Wikipedia: if you've ever looked up a gene on Wikipedia then chances are that you've seen work done as part of the Gene Wiki project. The Gene Wiki benefits from the many people who already dedicate time to editing and improving Wikipedia, and from the high visibility Wikipedia pages often have in search engines.

p 179: For another perspective on user-contributed comment sites for science, see [148].

p 179: The final report on *Nature*'s trial of open peer review: [167].

p 180: Although the user-contributed comment sites for science are failing, scientists aren't always unwilling to comment online about other scientists' work. We saw an example along these lines starting on page 259, with science bloggers investigating the evidence for chiropractic offered by the British Chiropractic Association in their dispute with Simon Singh. Other examples include (1) a Polymath-style collaboration [173] in 2010, in which a group of mathematicians, computer scientists, and physicists worked together online to analyze a claimed solution to one of the biggest open problems in computer science; (2) a blog-based online discussion [180] analyzing NASA's 2010 announcement [242] that they'd discovered lifeforms that incorporate arsenic; (3) Faculty of 1000 (http://f1000.com/), a site that actively recruits a limited number of high-profile researchers to write reviews of biomedical papers; and (4) MathSciNet (http://www.ams.org/mathscinet/), a similar site for mathematics. In each case, the incentives for potential contributors are quite different than for the user-contributed comment sites I have described. I won't analyze the incentives here — the point of this section isn't to comprehensively assay scientists' online commenting habits — but note that in each case a detailed analysis shows that the incentives for scientists to comment are much stronger than for the user-contributed comment sites.

p 182 "publish [papers] or perish," not "publish [data] or perish" is from [171].

Chapter 9. The Open Science Imperative

p 187: Tobias Osborne's research blog on quantum computing is at http://tjoresearchnotes.wordpress.com/. The idea of open notebook science has been developed in detail by Jean-Claude Bradley [26] and Cameron Neylon [147]. See also Bradley's blog (http://usefulchem.blogspot.com/) and Neylon's blog (http://cameronneylon.net/).

p 187 open science "would require most scientists to simultaneously and completely change their behaviour: [164].

p 188: Details about the Swedish change from driving on the left to driving on the right may be found in [217] and [97]. The language in my account is inspired by a wonderful sentence of Stephen Pinker [170], who wrote, "A switch from driving on the left to driving on the right could not begin with a daring nonconformist or a grass-roots movement but would have to be imposed from the top down (which is what happened in Sweden at 5 am, Sunday, September 3, 1967)."

p 188: In fact, the *Journal des Sçavans* has a claim to being the world's first scientific journal, as it began publication a couple of months before the *Philosophical Transactions of the Royal Society.* However, the point is debatable, as the *Journal des Sçavans* mixed scientific and nonscientific content.

p 188: Mary Boas Hall's comments about Oldenburg begging for information from the scientists of the day are given in [89] (page 159).

p 191: The policy situation for sharing of genetic data is rapidly evolving. For a broad overview of policy at the National Institutes of Health, including the policy on genome-wide association studies, see [143]. For the specific policy from the National Human Genome Research Institute in support of the Bermuda Agreement, see [96]. For the Wellcome Trust policy, see [236].

p 191: The UK Medical Research Council's policy on open data is in [229].

p 191 a spokesperson said this announcement was merely "phase one" of an effort to ensure that all data be openly accessible: [140].

p 191: The OECD recommendations on open access to publicly funded research data are in [160].

p 191: The phrase *"the republic of science"* is from Michael Polanyi's excellent essay [172] of the same title. Among other things, the essay describes the dangers of too much centralized control in science, exactly the sort of centralized control that grant agencies have today. (When Polanyi was writing, the grant agencies had far smaller budgets, and consequently much less power.) I agree with Polanyi's concerns — indeed, it's tempting to write a follow-up essay on "The Oligarchy of Science" — but the point of the current discussion is, of course, to find best actions in the world we find ourselves in, not in some idealized world.

p 193: On property rights in ideas and the invisible hand in science, see [172, 48]; an interesting general article on invisible hand explanations is [230]. I don't know where the term "reputation economy" originates; it has been in wide use since the 1990s (and perhaps earlier), but the idea is much older.

p 194: SPIRES is at http://www.slac.stanford.edu/spires/. The physics preprint arXiv is, as previously noted, at http://arxiv.org.

p 195: On new ways of measuring science, see, for instance, [175] and references therein.

p 196: Regarding the development of new tools for the construction of knowledge, I've placed most of the onus on scientists to build these tools. You might object that developing such tools is the job of academic libraries and scientific publishers. However, there are many reasons to think that the right place for such tools to originate is with scientists themselves. Consider, for example, that nearly all the examples I've described in this book — from the Polymath Project to GenBank to the arXiv — were created by scientists. Libraries and scientific publishers are not, for the most part, set up to work on such risky and radical innovations. Instead, they're oriented toward steady improvements to existing ways of doing things. While the libraries and publishers employ many talented people, when those people try to develop radically new tools they often find themselves battling tremendous institutional inertia. As a result, the best place for new tools to originate is with scientists themselves. I believe the appropriate role for libraries and publishers is later, as partners who can help sustain and further develop the most successful tools. This is exactly what has happened with projects such as the arXiv and GenBank, which were started by scientists, but whose growth and further development came through partnerships with the Cornell University Library and the US National Library of Medicine, respectively.

p 196: Continuing the theme of the last note, you might also wonder if developing new software tools might not be a job for a centralized agency. This has been tried in biology, for example, where many software tools are developed at the US's National Center for Biotechnology Information (NCBI), itself a part of the US National Library of Medicine. The NCBI is responsible for running GenBank and has also helped pioneer or support many other important online biological databases. But while the NCBI provides a valuable service, it also centralizes innovation, and drives out potential competitors, who cannot hope to compete with the deep pockets of the NCBI. Over the long run, I believe that science needs a more decentralized approach to innovation.

p 196: On the limits to measuring science, see [154].

p 198: On expectations about privacy, ethics, safety and legality, those expectations will, of course, evolve. Sites such as Patients Like Me (http://patientslikeme.com) ask medical patients to voluntarily share their medical information, and many patients have done so, in part so that the information can be used for research purposes.

p 198: The Grothendieck quote is from [85]. See also the discussion in chapter 18 of [200], where I learned of this quote.

p 198: The problem of managing attention in collaboration has been studied experimentally in [76]. Their results are consistent with the analysis here, and show that group problem solving may actually become less effective if everyone communicates with everyone else.

p 200: An account of the Trenberth email, together with a link to the (apparently genuine) original email, may be found in [44]. Trenberth's original paper [225] is quite readable.

p 201: On the management of the Kepler data, see [91, 166]. For the February 2011 announcement of Earth-size planets, see [129]. Note that in September 2010 another team independently announced [232] finding an Earth-like planet, around the star Gliese 581. This discovery has since been contested [110].

p 201: Dorigo's announcement that he was hearing rumors that the Higgs particle had been discovered is in [54], and his retraction is in [55]. Coverage in the mainstream media includes [47, 41].

p 202: A discussion of the history of the classification of the finite simple groups is given in [201]. The current status of the classification is discussed in [5].

p 203 How can other scientists verify and reproduce the results from such experiments?: See, e.g., [205], and references therein.

p 203: On "science beyond individual understanding," see [155].

p 203 Worldwide, our governments spend more than 100 billion dollars each year on basic research: I've based this assertion on chapter 4 of a report from the US National Science Foundation [144]. According to numbers included in that report, the US government spends 39 billion dollars each year on basic research. The report does not directly compute total worldwide governmental spending on basic research, and so the figure of 100 billion dollars is an estimate, based on several other numbers from that report.

p 206: The Daniel Hillis quote "there are problems that are impossible..." is from page 157 of Stewart Brand's book *The Clock of the Long Now* [27].

Appendix

p 211: A gentle introduction to the density Hales Jewett (DHJ) theorem, including on explanation of the concept of combinatorial lines, may be found in [66].

p 212: For Szemerédi's theorem, see [218]. The Green-Tao theorem is proved in [84].

p 212: The original proof of DHJ was in [66].

References

[1] Aotearoa. Comment on a blog post at *Boing Boing* (blog), May 3, 2009. http://www.boingboing.net/2009/05/03/wasting-time-for-a-g.html#comment-481275.

[2] Aotearoa. Comment on a blog post at *Nature newsblog*, March 23, 2009. http://blogs.nature.com/news/2008/05/addictive_protein_folding_game.html#comment-97818.

[3] Oliver Arafat and Dirk Riehle. The commit size distribution of open source software. *Proceedings of the 42nd Hawaiian International Conference on System Sciences*, 2008.

[4] G. Arnison et al. Experimental observation of lepton pairs of invariant mass around 95 GeV/c^2 at the CERN SPS collider. *Physics Letters B*, 126(5):398–410, 1983.

[5] Michael Aschbacher. The status of the classification of the finite simple groups. *Notices of the American Mathematical Society*, 51(7):736–740, August 2004. http://www.ams.org/notices/200407/fea-aschbacher.pdf.

[6] Nikolaos Askitas and Klaus F. Zimmermann. Google econometrics and unemployment forecasting. *Applied Economics Quarterly*, 55:107–120, 2009.

[7] Sitaram Asur and Bernardo A. Huberman. Predicting the future with social media. *eprint arXiv:1003.5699*, 2010.

[8] Audit Bureau of Circulations. ACCESS ABC: eCirc for US newspapers, 2010. http://abcas3.accessabc.com/ecirc/newstitlesearchus.asp.

[9] Robert M. Axelrod. *The Evolution of Cooperation*. New York: Basic Books, 1984.

[10] Manda Banerji, Ofer Lahav, Chris J. Lintott, Filipe B. Abdalla, Kevin Schawinski, Steven P. Bamford, Dan Andreescu, Phil Murray, M. Jordan Raddick, Anze Slosar, Alex Szalay, Daniel Thomas, and Jan Vandenberg. Galaxy Zoo: Reproducing galaxy morphologies via machine learning. *Monthly Notices of the Royal Astronomical Society*, 406(1):342–353, July 2010. eprint arXiv:0908.2033.

[11] J. Battelle. *The Search: How Google and Its Rivals Rewrote the Rules of Business and Transformed Our Culture*. Boston: Nicholas Brealey, 2005.

[12] Yochai Benkler. Coase's penguin, or, Linux and *The Nature of the Firm*. *The Yale Law Journal*, 112:369–446, 2002.

[13] Yochai Benkler. *The Wealth of Networks*. New Haven: Yale University Press, 2006.

[14] Tim Berners-Lee. *Weaving the Web*. New York: Harper Business, 2000.

[15] Tim Berners-Lee and James Hendler. Publishing on the semantic web. *Nature*, 410:1023–1024, April 26, 2001.

[16] Tim Berners-Lee, James Hendler, and Ora Lassila. The semantic web. *Scientific American*, May 17, 2001.

[17] Mario Biagioli. *Galileo's Instruments of credit: Telescopes, images, secrecy*. Chicago: University of Chicago Press, 2006.

[18] Peter Block. *Community: The Structure of Belonging*. San Francisco: Berrett Koehler, 2008.

[19] Barry Boehm, Bradford Clark, Ellis Horowitz, Ray Madachy, Richard Shelby, and Chris Westland. Cost models for future software life cycle processes: COCOMO 2.0. *Annals of Software Engineering*, 1(1):57–94, 1995.

[20] Peter Bogner, Ilaria Capua, David J. Lipman, and Nancy J. Cox et al. A global initiative on sharing avian flu data. *Nature*, 442:981, August 31, 2006.

[21] John Bohannon. Gamers unravel the secret life of protein. *Wired*, 17(5), April 20, 2009. http://www.wired.com/medtech/genetics/magazine/17-05/ff_protein?currentPage=all.

[22] Parsa Bonderson, Sankar Das Sarma, Michael Freedman, and Chetan Nayak. A blueprint for a topologically fault-tolerant quantum computer. *eprint arXiv:1003.2856*, 2010.

[23] Christine L. Borgman. *Scholarship in the Digital Age*. Cambrdge, MA: MIT Press, 2007.

[24] Kirk D. Borne et al. Astroinformatics: A 21st century approach to astronomy. *eprint arXiv: 0909.3892*, 2009. Position paper for Astro2010 Decadal Survey State, available at http://arxiv.org/abs/0909.3892.

[25] Todd A. Boroson and Tod R. Lauer. A candidate sub-parsec supermassive binary black hole system. *Nature*, 458:53–55, March 5, 2009.

[26] Jean-Claude Bradley. Open notebook science. *Drexel CoAS E-Learning* (blog), September 26, 2006. http://drexel-coas-elearning.blogspot.com/2006/09/open-notebook-science.html.

[27] Stewart Brand. *The Clock of the Long Now*. New York: Basic Books, 2000.

[28] John Seely Brown and Paul Duguid. *The Social Life of Information*. Boston: Harvard Business School Press, 2000.

[29] Zacary Brown. I'm a solver. *Perspectives on Innovation* (blog), February 4, 2009. http://blog.innocentive.com/2009/02/04/im-a-solver-zacary-brown/.

[30] Admiral Bumblebee. Comment on submission "Kasparov versus the World," 2007. http://www.reddit.com/r/reddit.com/comments/2hvex/kasparov_versus_the_world/.

[31] Vannevar Bush. As we may think. *Atlantic Monthly*, July 1945.

[32] Declan Butler. Flu database row escalates. *The Great Beyond* (blog), September 14, 2009. http://blogs.nature.com/news/thegreatbeyond/2009/09/flu_database_row_escalates.html.

[33] Robert H. Carlson. *Biology Is Technology*. Cambridge, MA: Harvard University Press, 2010.

[34] Nicholas Carr. Is Google making us stupid? *Atlantic Monthly*, July/August, 2008.

[35] Nicholas Carr. *The Shallows: What the Internet Is Doing to Our Brains.* New York: W. W. Norton & Company, 2010.

[36] Henry William Chesbrough. *Open Innovation: The new Imperative for Creating and Profiting from Technology.* Boston: Harvard Business Press, 2006.

[37] Chess Base. Dark horse ZackS wins Freestyle chess tournament, June 19, 2005. http://www.chessbase.com/newsdetail.asp?newsid=2461.

[38] Chess Base. Hydra misses the quarter-finals of Freestyle tournament, June 11, 2005. http://www.chessbase.com/newsdetail.asp?newsid=2446.

[39] Chess Base. PAL / CSS report from the dark horse's mouth, June 22, 2005. http://www.chessbase.com/newsdetail.asp?newsid=2467.

[40] The chess games of Hydra (Computer). http://www.chessgames.com/perl /chessplayer?pid=87303.

[41] Tom Chivers. Large Hadron Collider rival Tevatron "has found Higgs boson," say rumours. *Daily Telegraph*, July 12, 2010.

[42] Hyunyoung Choi and Hal Varian. Predicting the present with Google trends. *Google Research blog*, April 12, 2009. http://googleresearch.blogspot .com/2009/04/predicting-present-with-google-trends.html.

[43] Andy Clark and David J. Chalmers. The extended mind. *Analysis*, 58:10–23, 1998.

[44] "Climategate" exposed: Conservative media distort stolen emails in latest attack on global warming consensus. *Media Matters*, December 1, 2009. http://mediamatters.org/research/200912010002.

[45] Robert P. Colwell. *The Pentium Chronicles.* Hoboker, NJ: IEEE Computer Society, 2006.

[46] Seth Cooper, Firas Khatib, Adrien Treuille, Janos Barbero, Jeehyung Lee, Michael Beenen, Andrew Leaver-Fay, David Baker, Zoran Popović & Foldit players. Predicting protein structures with a multiplayer online game. *Nature*, 466:756–760, August 5, 2010.

[47] Rachel Courtland. Higgs boson: Is a result imminent? *New Scientist*, July 9, 2010.

[48] Partha Dasgupta and Paul A. David. Toward a new economics of science. *Research Policy*, 23:487–521, 1994.

[49] Paul A. David. The historical origins of "open science": An essay on patronage, reputation and common agency contracting in the scientific revolution. *Capitalism and Society*, 3(2), 2008.

[50] John R. Delaney and Roger S. Barga. A 2020 vision for ocean science. In Tony Hey, Stewart Tansley, and Kristin Tolle, editors, *The Fourth Paradigm: Data-Intensive Scientific Discovery.* Seattle: Microsoft Research, 2009. http://research.microsoft.com/en-us/collaboration/fourthparadigm/.

[51] Amit Deshpande and Dierk Riehle. The total growth of open source. In *Proceedings of the Fourth Conference on Open Source Systems*, 2008.

[52] David Diamond. The way we live now: Questions for Linus Torvalds. *New York Times*, September 28, 2003.

[53] Jared Diamond. *Collapse*. New York: Penguin Books, 2005.

[54] Tommaso Dorigo. Rumors about a light Higgs. *A Quantum Diaries Survivor* (blog), July 8, 2010. http://www.science20.com/quantum_diaries_survivor /rumors_about_light_higgs.

[55] Tommaso Dorigo. So was the rumor more than just a rumor, or was it a honest rumor? *A Quantum Diaries Survivor* (blog), July 17, 2010. http:// www.science20.com/quantum_diaries_survivor/so_was_rumor_more_just _rumor_or_was_it_honest_rumor.

[56] Robert Dougans and David Allen Green. Virtual veracity. *The Lawyer*, July 5, 2010.

[57] K. Eric Drexler. Hypertext publishing and the evolution of knowledge. *Social Intelligence*, 1:87–120, 1991.

[58] Jason Dyer. A gentle introduction to the Polymath Project. *The Number Warrior* (blog), March 25, 2009. http://numberwarrior.wordpress.com /2009/03/25/a-gentle-introduction-to-the-polymath-project/.

[59] David Easley and Jon Kleinberg. *Networks, Crowds, and Markets*. Cambridge: Cambridge University Press, 2010.

[60] Nature editorial. Dreams of flu data. *Nature*, 440:255–256, March 16, 2006.

[61] Elizabeth L. Eisenstein. *The Printing Revolution in Early Modern Europe (2nd ed.)*. Cambridge: Cambridge University Press, 2005.

[62] T. S. Eliot. *The Sacred Wood: Essays on Poetry and Criticism*. London: Methune, 1920.

[63] Douglas C. Engelbart. Augmenting human intellect: A conceptual framework. *Stanford Research Institute Report*, October 1962.

[64] Jon Fortt. Top 5 moments from Eric Schmidt's talk in Abu Dhabi. *Fortune Tech* (blog), March 11, 2010. http://brainstormtech.blogs.fortune.cnn.com /2010/03/11/top-five-moments-from-eric-schmidt%27s-talk-in-abu-dhabi/.

[65] Full cast and crew for Avatar. *Internet Movie Database (IMDb)*. http://www.imdb.com/title/tt0499549/fullcredits.

[66] Hillel Furstenberg and Yitzhak Katznelson. A density version of the Hales-Jewett theorem. *Journal d'Analyse Mathematique*, 57:64–119, 1991.

[67] Galaxy Zoo Forum. The Hanny's Voorwerp, 2007–∞. http://www.galaxy zooforum.org/index.php?topic=3802.0.

[68] Jeffrey S. Gerber and Paul A. Offit. Vaccines and autism: A tale of shifting hypotheses. *Clinical Infectious Diseases*, 48:456–461, 2009.

[69] Jim Giles. Google tops translation rankings. *Nature News*, November 7, 2006. http://www.nature.com/news/2006/061106/full/news061106-6.html.

[70] Jim Giles. PR's "pit bull" takes on open access. *Nature*, 445:347, February 1, 2007.

[71] Jeremy Ginsberg, Matthew H. Mohebbi, Rajan S. Patel, Lynnette Brammer, Mark S. Smolinski, and Larry Brilliant. Detecting influenza epidemics using search engine query data. *Nature*, 457:1012–1015, February 19, 2009.

[72] James Gleick. *Genius: The Life and Science of Richard Feynman*. Toronto: Random House of Canada, 1993.

[73] Sharad Goel, Jake M. Hofman, Sébastien Lahaie, David M. Pennock, and Duncan J. Watts. What can search predict? http://www.cam.cornell.edu/~sharad/papers/searchpreds.pdf, 2009.

[74] Ben Goldacre. An intrepid, ragged band of bloggers. *Guardian*, July 29, 2009. http://www.badscience.net/2009/07/we-are-more-possible-than-you-can-powerfully-imagine/.

[75] Michael H. Goldhaber. The attention economy and the net. *First Monday*, 2(4–7), April 1997.

[76] Robert L. Goldstone, Michael E. Roberts, and Todd M. Gureckis. Emergent processes in group behavior. *Current Directions in Psychological Science*, 17(1):10–15, 2008.

[77] J. Richard Gott III, Mario Jurić, David Schlegel, Fiona Hoyle, Michael Vogeley, Max Tegmark, Neta Bahcall, and Jon Brinkmann. A map of the universe. *Astrophysical Journal*, 624(2):463–484, 2005.

[78] W. Timothy Gowers. Comment on *Gowers's weblog*, February 2, 2009. http://gowers.wordpress.com/2009/02/01/questions-of-procedure/#comment-1701.

[79] W. Timothy Gowers. Is massively collaborative mathematics possible? *Gowers's weblog*, January 27, 2009. http://gowers.wordpress.com/2009/01/27/is-massively-collaborative-mathematics-possible/.

[80] W. Timothy Gowers. Polymath1 and open collaborative mathematics. *Gowers's weblog*, March 10, 2009. http://gowers.wordpress.com/2009/03/10/polymath1-and-open-collaborative-mathematics/.

[81] W. Timothy Gowers. Problem solved (probably). *Gowers's weblog*, March 10, 2009. http://gowers.wordpress.com/2009/03/10/problem-solved-probably/.

[82] W. Timothy Gowers and Michael Nielsen, Massively collaborative mathematics, *Nature* 461, October 15, 2009.

[83] Jim Gray. eScience: A transformed scientific method. In Tony Hey, Stewart Tansley, and Kristin Tolle, editors, *The Fourth Paradigm: Data-Intensive Scientific Discovery*. Seattle: Microsoft Research, 2009. http://research.microsoft.com/en-us/collaboration/fourthparadigm/.

[84] Ben Green and Terence Tao. The primes contain arbitrarily long arithmetic progressions. *Annals of Mathematics*, 167:481–547, 2008.

[85] Alexander Grothendieck. *Recoltes et Semailles*. 1986. http://www.ferment magazine.org/rands/recoltes1.html.

[86] Ned Gulley. In praise of tweaking: A wiki-like programming contest. *Interactions*, 11(3):18–23, May–June 2004.

[87] Ned Gulley and Karim R. Lakhani. The determinants of individual performance and collective value in private-collective software innovation. Harvard Business School Working Paper 10–65, 2010.

[88] Alon Halevy, Peter Norvig, and Fernando Pereira. The unreasonable effectiveness of data. *IEEE Intelligent Systems*, 24:8–12, 2009.

[89] Marie Boas Hall. *Henry Oldenburg: Shaping the Royal Society*. Oxford: Oxford University Press, 2002.

[90] Michael J. Hammel. Industry of change: Linux storms Hollywood. *Linux Journal*, February 2002. http://www.linuxjournal.com/article/5472.

[91] Eric Hand. Telescope team may be allowed to sit on exoplanet data. *Nature News*, April 14, 2010. http://www.nature.com/news/2010/100414/full/news.2010.182.html.

[92] Chuck Hansen. *U.S. Nuclear Weapons: The Secret History*. Arlington, Aerofax, 1988.

[93] Friedrich von Hayek. The use of knowledge in society. *American Economic Review*, 35(4):519–530, 1945.

[94] Tony Hey, Stewart Tansley, and Kristin Tolle, editors. *The Fourth Paradigm: Data-Intensive Scientific Discovery*. Seattle: Microsoft Research, 2009. http://research.microsoft.com/en-us/collaboration/fourthparadigm/.

[95] Edwin Hutchins. *Cognition in the Wild*. Cambridge, MA: MIT Press, 1995.

[96] National Human Genome Research Institute. Reaffirmation and extension of NHGRI rapid data release policies: Large-scale sequencing and other community resource projects, February 2003. http://www.genome.gov/10506537.

[97] In one instant a left-lane nation swerves right. *Life*, September 15, 1967.

[98] Jane Jacobs. *The Death and Life of Great American Cities*. Toronto: Random House of Canada, 1961.

[99] Irving Lester Janis. *Groupthink: Psychological Studies of Policy Decisions and Fiascoes*. Boston: Houghton Mifflin, 1983.

[100] Eamon Javers. The pit bull of public relations. *Business Week*, April 17, 2006.

[101] Ayodele Samuel Jegede. What led to the Nigerian boycott of the polio vaccination campaign? *PLoS Medicine*, 4(3):e73, 2007. http://www.plosmedicine.org/article/info:doi/10.1371/journal.pmed.0040073.

[102] Joint statement by President Clinton and Prime Minister Blair. March 14, 2000. http://clinton4.nara.gov/WH/EOP/OSTP/html/00314.html.

[103] Joint United Nations Programme on HIV/AIDS. Socio-economic impact of the epidemic and the strengthening of national capacities to combat HIV/AIDS, June 15, 2001. http://data.unaids.org/Publications/External-Documents/GAS26-rt3_en.pdf.

[104] Jonathan Spencer Jones. J. C. Bennett (1914–1990). *Quarterly Journal of the Royal Astronomical Society*, 35:353, 1994.

[105] G.I.G. Jozsa, M. A. Garrett, T. A. Oosterloo, H. Rampadarath, Z. Paragi, H. van Arkel, C. Lintott, W. C. Keel, K. Schawinski, and E. Edmondson. Revealing Hanny's Voorwerp: Radio observations of IC 2497. *Astronomy and Astrophysics*, 500(2):L33–L36, 2009. eprint arXiv:0905.1851.

[106] Garry Kasparov. The chess master and the computer. *New York Review of Books*, 57(2), February 11, 2010.

[107] Garry Kasparov with Daniel King. *Kasparov Against the World*. KasparovChess Online, 2000.

[108] Kevin Kelly. Speculations on the future of science. *Edge: The Third Culture*, 2006. http://www.edge.org/3rd_culture/kelly06/kelly06_index.html.

[109] Kevin Kelly. *What Technology Wants*. New York: Viking, 2010.

[110] Richard A. Kerr. Recently discovered habitable world may not exist. *Science Now*, October 12, 2010. http://news.sciencemag.org/sciencenow/2010/10 /recently-discovered-habitable-world.html.

[111] A. Yu Kitaev. Fault-tolerant quantum computation by anyons. *Annals of Physics*, 303(1):2–30, 2003.

[112] Helge Kragh. Max Planck: The reluctant revolutionary. *Physics World*, December 2000. http://physicsworld.com/cws/article/print/373.

[113] Greg Kroah-Hartman. The Linux kernel. Online video from Google Tech Talks. http://www.youtube.com/watch?v=L2SED6sewRw.

[114] Greg Kroah-Hartman, Jonathan Corbet, and Amanda McPherson. Linux kernel development. *The Linux Foundation*, April 2008.

[115] Irina Krush with Kenneth W. Regan. The greatest game in the history of chess, parts I, II, and III. Available at http://www.cse.buffalo.edu/~regan /chess/K-W/KHR99i.html, 1999.

[116] Karim R. Lakhani, Lars Bo Jeppesen, Peter A. Lohse, and Jill A. Panetta. The value of openness in scientific problem solving. Harvard Business School Working Paper 07-050, 2007.

[117] Jaron Lanier. *You Are Not a Gadget: A Manifesto*. Toronto: Random House of Canada, 2010.

[118] Hadley Leggett. Aug. 18, 1868: Helium discovered during total solar eclipse. *Wired*, August 18, 2009. http://www.wired.com/thisdayintech/2009 /08/dayintech_0818/.

[119] Jonah Lehrer. Making connections. *Nature*, 457:524–527, January 28, 2009.

[120] Jonah Lehrer. Scientists map the brain, gene by gene. *Wired*, 17, March 28, 2009. http://www.wired.com/medtech/health/magazine/17-04/ff _brainatlas.

[121] Ed S. Lein *et al.* Genome-wide atlas of gene expression in the adult mouse brain. *Nature*, 445:168–176, January 11, 2007.

[122] Lawrence Lessig. *Free Culture: How Big Media Uses Technology and the Law to Lock Down Culture and Control Creativity*. New York: Penguin, 2004. http://www.free-culture.cc/freecontent/.

[123] Pierre Lévy. *L'intelligence collective*. Paris: La Découverte, 1994.

[124] Pierre Lévy. *Collective Intelligence*. Cambridge, MA: Perseus Books, 1997. Translated from the French original [123] by Robert Bononno.

[125] Jeff W. Lichtman, R. Clay Reid, Hanspeter Pfister, and Michael F. Cohen. Discovering the wiring diagram of the brain. In Tony Hey, Stewart Tansley, and Kristin Tolle, editors, *The Fourth Paradigm: Data-Intensive Scientific Discovery*. Seattle: Microsoft Research, 2009. http://research.microsoft.com /en-us/collaboration/fourthparadigm/.

[126] Chris Lintott. He said that they said that he said.... *Galaxy Zoo* (blog), 2009. http://blogs.zooniverse.org/galaxyzoo/2009/07/09/he-said-that-they -said-that-he-said/.

[127] Chris Lintott and Kate Land. Eyeballing the universe. *Physics World*, 21:27– 30, 2008.

[128] Chris J. Lintott, Kevin Schawinski, William Keel, Hanny van Arkel, Nicola Bennert, Edward Edmondson, Daniel Thomas, Daniel J. B. Smith, Peter D. Herbert, Matt J. Jarvis, Shanil Virani, Dan Andreescu, Steven P. Bamford, Kate Land, Phil Murray, Robert C. Nichol, M. Jordan Raddick, Anže Slosar, Alex Szalay, and Jan Vandenberg. Galaxy Zoo: "Hanny's Voorwerp", a quasar light echo? *Monthly Notices of the Royal Astronomical Society*, 399(1):129– 140, October 2009. eprint arXiv:0906.5304.

[129] Jack J. Lissauer, Daniel C. Fabrycky, Eric B. Ford, William J. Borucki, Fran- cois Fressin, Geoffrey W. Marcy, Jerome A. Orosz, Jason F. Rowe, Guillermo Torres, William F. Welsh, Natalie M. Batalha, Stephen T. Bryson, Lars A. Buchhave, Douglas A. Caldwell, Joshua A. Carter, David Charbonneau, Jessie L. Christiansen, William D. Cochran, Jean-Michel Desert, Edward W. Dunham, Michael N. Fanelli, Jonathan J. Fortney, Thomas N. Gautier III, John C. Geary, Ronald L. Gilliland, Michael R. Haas, Jennifer R. Hall, Matthew J. Holman, David G. Koch, David W. Latham, Eric Lopez, Sean McCauliff, Neil Miller, Robert C. Morehead, Elisa V. Quintana, Darin Ragozzine, Dimitar Sasselov, Donald R. Short, and Jason H. Steffen. A closely packed system of low-mass, low-density planets transiting Kepler-11. *Nature*, 470:53–58, February 3, 2011.

[130] Charles Mackay. *Extraordinary popular delusions and the madness of crowds* (1841). Toronto: Random House of Canada, 1995.

[131] Emma Marris. American Chemical Society: Chemical reaction. *Nature*, 437:807–809, October 6, 2005.

[132] Karen Masters. She's an astronomer: Aida Berges. *Galaxy Zoo* (blog), October 1, 2009. http://blogs.zooniverse.org/galaxyzoo/2009/10/01/shes-an -astronomer-aida-berges/.

[133] Hassan Masum and Mark Tovey. Given enough minds . . . : Bridging the ingenuity gap. *First Monday* 11 (7), July 2006.

[134] Hassan Masum and Mark Tovey, editors. *The Reputation Society*. Cam- bridge, MA: MIT Press, forthcoming.

[135] Peter McIntyre and Julie Leask. Improving uptake of MMR vaccine. *British Medical Journal*, 336:729–730, 2008.

[136] Lucas Mearian. CDC adopts new, near real-time flu tracking system. *Com- puter World*, November 5, 2009.

[137] Meatball Wiki. WikiAsScience. http://meatballwiki.org/wiki/WikiAsScience.

[138] Meatball Wiki. WikiSciencePublication. http://meatballwiki.org/wiki/Wiki SciencePublication.

[139] David Mehegan. Author(s)! author(s)! *Off the Shelf* (blog), April 10, 2007. http://www.boston.com/ae/books/blog/2007/04/authors_authors_2.html.

[140] Jeffrey Mervis. NSF to ask every grant applicant for data management plan. *ScienceInsider*, May 5, 2010. http://news.sciencemag.org/scienceinsider /2010/05/nsf-to-ask-every-grant-applicant.html.

[141] Morbidity and mortality weekly report. United States Center for Disease Control, October 13, 2006.

[142] Craig Mundie. The commercial software model. Speech at the New York University Stern School of Business, May 2001. http://www.microsoft.com /presspass/exec/craig/05-03sharedSource.mspx.

[143] National Institutes of Health. NIH data sharing policy (as of April 17, 2007). http://grants.nih.gov/grants/policy/data_sharing/.

[144] National Science Foundation. Science and engineering indicators, 2010. http://www.nsf.gov/statistics/seind10/pdfstart.htm.

[145] Theodor Holm Nelson. *Literary Machines*. Sausalito, CA: Mindful Press, 1987.

[146] Cameron Neylon. The science exchange. *Science in the Open* (blog), April 16, 2008. http://cameronneylon.net/blog/the-science-exchange/.

[147] Cameron Neylon. Scientists lead the push for open data sharing. *Research Information*, April/May 2009. http://www.researchinformation.info/features /feature.php?feature_id=214.

[148] Cameron Neylon and Shirley Wu. Article-level metrics and the evolution of scientific impact. *PLoS Biology* 7(11): e1000242, 2009.

[149] Bob Nichol. This is my first time.... *Galaxy Zoo* (blog), February 19, 2009. http://blogs.zooniverse.org/galaxyzoo/2009/02/19/this-is-my-first-time/.

[150] Michael Nielsen. Doing science in the open. *Physics World*, May 2009. http://physicsworld.com/cws/article/print/38904.

[151] Michael Nielsen. The economics of scientific collaboration. *Michael Nielsen's blog*, December 29, 2008. http://michaelnielsen.org/blog/the-economics-of -scientific-collaboration/.

[152] Michael Nielsen. The future of science. *Michael Nielsen's blog*, July 17, 2008. http://michaelnielsen.org/blog/the-future-of-science-2/.

[153] Michael Nielsen, Information awakening, Nature Physics 5, April 2009.

[154] Michael Nielsen. The mismeasurement of science. *Michael Nielsen's blog*, November 29, 2010. http://michaelnielsen.org/blog/the-mismeasurement -of-science/, and to appear in [134].

[155] Michael Nielsen. Science beyond individual understanding. *Michael Nielsen's blog*, September 24, 2008. http://michaelnielsen.org/blog/science-beyond -individual-understanding/.

[156] Nielsen Media Research. Three screen report. *Nielsen wire* (blog), Q1 2009. http://blog.nielsen.com/ nielsenwire/online_mobile/americans-watching -more-tv-than-ever/.

[157] Peter Norvig. How to write a spelling corrector. http://norvig.com/spell -correct.html.

[158] Barack Obama. Transparency and open government. *Federal Register*, 74(15), January 26, 2009. http://www.whitehouse.gov/the_press_office/Transparency andOpenGovernment/.

[159] Ryan O'Donnell. Comment on *Gowers's weblog*, February 6, 2009. http://gowers.wordpress.com/2009/02/06/dhj-the-triangle-removal-approach/#comment-1913.

[160] OECD. OECD principles and guidelines for access to research data from public funding. OECD Report, April 2007. http://www.oecd.org/document /55/0,3343,en_2649_201185_38500791_1_1_1_1,00.html.

[161] Mancur Olson. *The Logic of Collective Action: Public Goods and the Theory of Groups*. Cambridge, MA: Harvard University Press, 1965.

[162] Tim O'Reilly. The architecture of participation, June 2004. http://www.oreillynet.com/pub/a/oreilly/tim/articles/architecture_of_participation .html.

[163] Justin R. Ortiz, Hong Zhou, David K. Shay, Kathleen M. Neuzil, and Christopher H. Goss. Does Google influenza tracking correlate with laboratory tests positive for influenza? Conference abstract. *American Journal of Respiratory and Critical Care Medicine*, 181:A2626, 2010.

[164] Tobias J. Osborne. Over 6 months later. *Tobias J. Osborne's research notes* (blog), October 4, 2010. http://tjoresearchnotes.wordpress.com/2009/10/04 /over-6-months-later/.

[165] Elinor Ostrom. *Governing the Commons: The Evolution of Institutions for Collective Action*. Cambridge: Cambridge University Press, 1990.

[166] Dennis Overbye. In the hunt for planets, who owns the data? *New York Times*, June 14, 2010. http://www.nytimes.com/2010/06/15/science/space /15kepler.html.

[167] Overview: Nature's peer review trial. *Nature*, December 2006. http://www .nature.com/nature/peerreview/debate/nature05535.html.

[168] Scott E. Page. *The Difference: How the Power of Diversity Creates Better Groups*. Princeton, NJ: Princeton University Press, 2008.

[169] A. Pais. *Subtle Is the Lord: The Science and the Life of Albert Einstein*. Oxford: Oxford University Press, 1982.

[170] Stephen Pinker. *The Blank Slate: The Modern Denial of Human Nature*. New York: Penguin, 2003.

[171] Elizabeth Pisani and Carla AbouZahr. Sharing health data: Good intentions are not enough. *Bulletin of the World Health Organization*, 88(6), 2010.

[172] Michael Polanyi. The republic of science: Its political and economic theory. *Minerva*, 1:54–74, 1962. http://www.missouriwestern.edu/orgs/polanyi/mp -repsc.htm.

[173] Polymath participants. Deolalikar P vs NP paper. *Polymath wiki*, 2010–. http://michaelnielsen.org/polymath1/index.php?title=Deolalikar's_P!%3DNP _paper.

[174] Zoran Popović. CASP8 results. *Foldit blog*, December 17, 2008. http://fold.it/portal/node/729520.

[175] Jason Priem, Dario Taraborelli, Paul Groth, and Cameron Neylon. Altmetrics: A manifesto. October 26, 2010. http://altmetrics.org/manifesto/.

[176] PRISM: Current issues. http://web.archive.org/web/20080330235026/http://www.prismcoalition.org/topics.htm.

[177] H. Rampadarath, M. A. Garrett, G. I. G. Józsa, T. Muxlow, T. A. Oosterloo, Z. Paragi, R. Beswick, H. van Arkel, W. C. Keel, and K. Schawinski. Hanny's Voorwerp: Evidence of AGN activity and a nuclear starburst in the central regions of IC 2497. *eprint arXiv:1006.4096*, 2010.

[178] Eric S. Raymond. The Cathedral and the Bazaar. Published online and reprinted in [179]. http://www.catb.org/~esr/writings/cathedral-bazaar/cathedral-bazaar/.

[179] Eric S. Raymond. *The Cathedral and the Bazaar: Musings on Linux and Open Source by an Accidental Revolutionary*. Sebastopol, CA: O'Reilly Media, 2001.

[180] Rosie Redfield. Arsenic-associated bacteria (NASA's claims). *Rrresearch* (blog), December 4, 2010. http://rrresearch.blogspot.com/2010/12/arsenic-associated-bacteria-nasas.html.

[181] Reed Elsevier. Annual reports and financial statements, 2009. http://reports.reedelsevier.com/ar09/

[182] Richard Rhodes. *How to Write*. New York: Harper Collins, 1995.

[183] Richard Rhodes. *The Making of the Atomic Bomb*. New York: Simon & Schuster, 1986.

[184] Ben Rich. *Skunk Works: A Personal Memoir of My Years of Lockheed*. Boston: Little, Brown and Company, 1996.

[185] Martin Robbins. A review of the BCA's evidence for chiropractic. *The Lay Scientist: Martin's blog*, 2009. http://www.layscience.net/node/598.

[186] Bill Rosato. Chess champion Kasparov meets match on internet. *Reuters (London)*, September 3, 1999.

[187] Robin Rowe. Linux #1 operating system in Hollywood. http://www.linuxmovies.org, 2008.

[188] Donald Rumsfeld. United States Department of Defense News Briefing, February 12, 2002. http://www.defense.gov/transcripts/transcript.aspx?transcriptid=2636.

[189] Thomas C. Schelling. *Micromotives and Macrobehavior*. New York: W. W. Norton & Company, 1978.

[190] Paul Seabright. *The Company of Strangers: A Natural History of Economic Life*. Princeton, NJ: Princeton University Press, 2004.

[191] Toby Segaran. *Programming Collective Intelligence*. Sebastopol, CA: O'Reilly Media, 2007.

[192] D. Shasha and C. Lazere. *Out of Their Minds: The Lives and Discoveries of 15 Great Computer Scientists*. New York: Springer-Verlag, 1998.

[193] Alice Sheppard. Peas in the universe, goodwill and a history of Zooite collaboration on the peas project. *Galaxy Zoo* (blog), July 7, 2009. http://blogs.zooniverse.org/galaxyzoo/2009/07/07/peas-in-the-universe-goodwill-and-a-history-of-zooite-collaboration-on-the-peas-project/.

[194] Clay Shirky. *Cognitive surplus: Creativity and generosity in a connected age*. Penguin, 2010.

[195] Clay Shirky. Gin, television, and social surplus. *Here Comes Everybody* (blog), April 26, 2008. http://www.shirky.com/herecomeseverybody/2008/04/looking-for-the-mouse.html.

[196] Clay Shirky. *Here comes everybody: The power of organizing without organizations.* New York: Penguin, 2008.

[197] Herbert A. Simon. Designing organizations for an information-rich world. In Martin Greenberger, editor, *Computers, Communication, and the Public Interest.* Baltimore: Johns Hopkins Press, 1971.

[198] Cameron Sinclair. Cameron Sinclair on open-source architecture. *TED: Ideas Worth Spreading,* 2006. http://www.ted.com/talks/cameron_sinclair _on_open_source_architecture.html.

[199] Simon Singh. Beware the spinal trap. *Guardian,* April 19, 2008.

[200] Lee Smolin. *The Trouble with Physics.* London: Allen Lane, 2006.

[201] Ron Solomon. On finite simple groups and their classification. *Notices of the American Mathematical Society,* 42(2):231–239, February 1995. http://www.ams.org/notices/199502/solomon.pdf.

[202] Richard M. Stallman. *Free Software, Free Society: Selected Essays of Richard M. Stallman.* Boston: Free Software Foundation, 2002. http://www.gnu.org /philosophy/fsfs/rms-essays.pdf.

[203] Garol Stasser and William Titus. Hidden profiles: A brief history. *Psychological Inquiry,* 14(3&4):304–313, 2003.

[204] Garold Stasser and William Titus. Pooling of unshared information in group decision making: Biased information sampling during discussion. *Journal of Personality and Social Psychology,* 48(6):1467–1478, 1985.

[205] Victoria Stodden, David Donoho, Sergey Fomel, Michael P. Friedlander, Mark Gerstein, Randy LeVeque, Ian Mitchell, Lisa Larrimore Ouellette, Chris Wiggins, Nicholas W. Bramble, Patrick O. Brown, Vincent J. Carey, Laura DeNardis, Robert Gentleman, J. Daniel Gezelter, Alyssa Goodman, Matthew G. Knepley, Joy E. Moore, Frank A. Pasquale, Joshua Rolnick, Michael Seringhaus, and Ramesh Subramanian. Reproducible research: Addressing the need for data and code sharing in computational science. *Computing in Science and Engineering,* p 12(5):8–12, Sep/Oct 2010.

[206] Peter Suber. A day worth celebrating. *Open Access News* (blog), April 17, 2008. http://www.earlham.edu/~peters/fos/2008/04/day-worth-celebrating. html.

[207] Peter Suber. Open access overview. http://www.earlham.edu/~peters/fos /overview.htm.

[208] Peter Suber. Timeline of the open access movement. http://www.earlham. edu/~peters/fos/timeline.htm.

[209] Brian Sullivan. Do you eBird?—open thread. *Chip Notes: eBird Buzz* (blog), May 23, 2009. http://ebirdforum.blogspot.com/2009/05/do-you-ebird-tell -us-about-yourself.html.

[210] Brian L. Sullivan, Christopher L. Wood, Marshall J. Iliff, Rick E. Bonney, Daniel Finka, and Steve Kellinga. eBird: A citizen-based bird observation

network in the biological sciences. *Biological Conservation*, 142(10):2282–2292, 2009.

[211] John Sulston. Heritage of humanity. *Le Monde Diplomatique (English Edition)*, November 2002.

[212] Cass R. Sunstein. *Infotopia: How Many Minds Produce Knowledge*. New York: Oxford University Press, 2006.

[213] Cass R. Sunstein. *Republic.com 2.0*. Princeton University Press, 2007.

[214] James Surowiecki. *The Wisdom of Crowds*. New York: Doubleday, 2004.

[215] Don R. Swanson. Migraine and magnesium: Eleven neglected connections. *Perspectives in Biology and Medicine*, 31(4):526–557, 1988.

[216] Don R. Swanson. Medical literature as a potential source of new knowledge. *Bulletin of the Medical Library Association*, 78(1):29–37, 1990.

[217] Sweden: Switch to the right. *Time*, September 15, 1967.

[218] Endre Szemerédi. On sets of integers containing no k elements in arithmetic progression. *Acta Arithmetica*, 27:199–245, 1975.

[219] Jeffery K. Taubenberger and David M. Morens. 1918 Influenza: The Mother of All Pandemics. *Emerging Infectious Diseases*, 12:15–22, 2006.

[220] Michael P. Taylor, Andrew A. Farke, and Mathew J. Wede. The open dinosaur project. *Palaeontological Association Newsletter*, (73), 2010. http://opendino.wordpress.com/2010/04/22/new-odp-article-in-the-palaeontological-association-newsletter/.

[221] Third update on BCA v Simon Singh, June 2009. http://www.chiropractic-uk.co.uk/gfx/uploads/textbox/Singh/BCA%20Statement%20170609.pdf.

[222] 31-year-old Texas native develops solar-powered wireless router for ASSET India, a non-profit organization focused on educating marginalized children in India in technology. *InnoCentive Press Release*, September 25, 2008. http://www.marketwire.com/press-release/31-Year-Old-Texas-Native-Develops-Solar-Powered-Wireless-Router-ASSET-India-Non-Profit-903974.htm.

[223] Linus Torvalds. The Linux edge. In *Open Sources: Voices from the Open Source Revolution*. editors, Chris DiBona, Sam Ockman, and Mark Stone, Sebastopol, CA: O'Reilly Media, 1999.

[224] Mark Tovey, editor. *Collective Intelligence: Creating a Prosperous World at Peace*. Oakton, VA: Earth Intelligence Network, 2008.

[225] Kevin Trenberth. An imperative for climate change planning: Tracking Earth's global energy. *Current Opinion in Environmental Sustainability*, 1:19–27, 2009. http://www.cgd.ucar.edu/cas/Trenberth/trenberth.papers/EnergyDiagnostics09final2.pdf.

[226] Ilkka Tuomi. Evolution of the Linux Credits file: Methodological challenges and reference data for open source research. *First Monday*, 9(6), 2004.

[227] Jon Udell. Internet groupware for scientific collaboration. 2000. http://jonudell.net/GroupwareReport.html.

[228] Jon Udell. Sam's encounter with manufactured serendipity. *Jon Udell's Radio blog*, March 4, 2002. http://radio-weblogs.com/0100887/2002/03/04.html.

[229] UK Medical Research Council policy on data sharing and preservation. http://www.mrc.ac.uk/Ourresearch/Ethicsresearchguidance/Datasharingini tiative/Policy/index.htm.

[230] Edna Ullmann-Margalit. Invisible-hand explanations. *Synthese*, 39(2): 263–291, 1978.

[231] Vernor Vinge. *Rainbows End*. New York: Tor, 2007.

[232] Steven S. Vogt, R. Paul Butler, Eugenio J. Rivera, Nader Haghighipour, Gregory W. Henry, and Michael H. Williamson. The Lick-Carnegie Exoplanet Survey: A 3.1 M_Earth planet in the habitable zone of the nearby M3V star Gliese 581. *eprint arXiv:1009.5733*, 2010.

[233] Eric von Hippel. *Democratizing Innovation*. Cambridge, MA: MIT Press, 2005.

[234] James D. Watson. *The Double Helix: A Personal Account of the Discovery of the Structure of DNA*. New York: Simon and Schuster, 1980.

[235] Steven Weber. *The Success of Open Source*. Cambridge, MA: Harvard University Press, 2004.

[236] Wellcome Trust. Policy on data management and sharing. 2010. http:// www.wellcome.ac.uk/About-us/Policy/Policy-and-position-statements/ WTX035043.htm.

[237] Wellcome Trust. Sharing data from large-scale biological research projects: A system of tripartite responsibility. 2003. http://www.wellcome .ac.uk/stellent/groups/corporatesite/@policy_communications/documents/web _document/wtd003207.pdf.

[238] Richard S. Westfall. Science and patronage: Galileo and the telescope. *Isis*, 76:11–30, 1985.

[239] What is SourceForge.net? http://sourceforge.net/apps/trac/sourceforge/wiki /What%20is%20SourceForge.net?

[240] John A. Wheeler. *A Journey into Gravity and Spacetime*. New York: Scientific American Library, 1990.

[241] John Willinsky. *The Access Principle*. The MIT Press, Cambridge, Massachusetts, 2006.

[242] Felisa Wolfe-Simon, Jodi Switzer Blum, Thomas R. Kulp, Gwyneth W. Gordon, Shelley E. Hoeft, Jennifer Pett-Ridge, John F. Stolz, Samuel M. Webb, Peter K. Weber, Paul C. W. Davies, Ariel D. Anbar, and Ronald S. Oremland. A bacterium that can grow by using arsenic instead of phosphorus. *Science*, December 2, 2010.

[243] Anita Williams Woolley, Christopher F. Chabris, Alex Pentland, Nada Hashmi, and Thomas W. Malone. Evidence for a collective intelligence factor in the performance of human groups. *Science*, 330(6004):686–688, October 29, 2010.

[244] World Health Organization. Influenza fact sheet number 211, March 2003. http://www.who.int/mediacentre/factsheets/2003/fs211/en/.

[245] Lynn Wu and Erik Brynjolfsson. The future of prediction: How Google searches foreshadow housing prices and sales. Presented at the 2009

Workshop on Information Systems and Economics (WISE 2009), 2009. http://pages.stern.nyu.edu/~bakos/wise/papers/wise2009-3b3_paper.pdf.

[246] Shirley Wu. Envisioning the scientific community as One Big Lab. *One Big Lab* (blog), April 14, 2008. http://onebiglab.blogspot.com/2008 /04/envisioning-scientific-community-as-one.html.

[247] Donald G. York *et al.* The Sloan digital sky survey: Technical summary. *Astronomical Journal*, 120(3):1579–1587, September 2000.

Index